MOLECULAR BIOLOGY OF EUCARYOTIC CELLS

A Problems Approach

Volume I

Leroy E. Hood
California Institute of Technology

John H. Wilson
Baylor College of Medicine

William B. Wood
California Institute of Technology

W. A. Benjamin, Inc.
Menlo Park, California · Reading, Massachusetts
London · Amsterdam · Don Mills, Ontario · Sydney

Other books by these authors:

BIOCHEMISTRY
A Problems Approach (©1974)

Wood, Wilson, Benbow, Hood

MOLECULAR BIOLOGY OF EUCARYOTIC CELLS
A Problems Approach, Volume II *(forthcoming)*

Hood, Wilson, Wood

ISBN 0-8053-9851-1
BCDEFGHIJ-AL-798765

W. A. Benjamin, Inc.
2725 Sand Hill Road
Menlo Park, California 94025

PREFACE

These books* draw together material from a number of currently exciting areas of molecular and cellular biology in a form that we hope will be of value to students and professionals alike. The central theme of the book is the structure and function of eucaryotic chromosomes as the vehicles of gene expression during development, gene transmission between generations, and gene evolution. Current understanding in these areas has come from a variety of biochemical, genetic, and ultrastructural approaches. Consequently, the book includes chapters on the relatively well understood organization and control of genes in procaryotes, the structure of eucaryotic chromosomes, and genetic analysis of eucaryotic organisms as well as their component somatic cells in culture. Gene expression is considered in the general context of several developmental systems, and specifically with regard to its control by hormones. Finally, evolution is considered from the standpoint of how genes and proteins change with time, and how these changes become fixed in populations. Throughout, the emphasis is placed on current attempts to understand these processes at the molecular level.

The book originally was undertaken as an introduction to eucaryotic molecular and cellular biology for biochemistry students. Consequently, it is organized according to the problems approach used in its companion volume,† in the belief that one of the best ways to learn this material is by actively grappling with experimental questions and analyzing experimental data. Each chapter is divided into four sections: Essential Concepts, References, Problems, and Answers.

The Essential Concepts sections are intended to concisely present important aspects of each topic. Additional concepts are introduced in the Problems, as indicated at the end of each Essential Concepts section. Three of the chapters also include appendices, which consider in step by step detail the rationale for three important contemporary analytical techniques: DNA-DNA reassociation, DNA-RNA hybridization, and the construction of genealogical trees from macromolecular sequence data.

The References include short "Where to begin" sections of readable introductory materials, followed by longer lists of both general and specific articles from the literature. An effort has been made to include classic papers as well as the most recent references.

The Problems sections include from 16 to 29 exercises of increasing difficulty. Many problems have been drawn from the contemporary literature; the reader is given data, real or simulated, from a variety of genetic, biochemical,

---

*L. E. Hood, J. H. Wilson, and W. B. Wood, *Molecular Biology of Eucaryotic Cells: A Problems Approach. Volume II.* In preparation (to include chapters on antibodies and the immune system, cell surfaces, cyto-architecture, animal viruses, and cancer).

†W. B. Wood, J. H. Wilson, R. M. Benbow, and L. E. Hood, *Biochemistry: A Problems Approach,* W. A. Benjamin, Menlo Park, California, 1974. This book will be referred to throughout this volume as WWBH.

and biophysical experiments, and is asked to analyze them as did the original investigators. All the information required to work these problems is contained in the book.

The Answers sections provide the reader with detailed feedback on his efforts to obtain solutions, often by describing the analytical process used originally to interpret data from the literature. It should be noted that many of the problems may be solved by alternative approaches, although generally only a single approach is explained in the answer. Answers to more challenging problems are written so that the first sentence or two will provide a hint toward the solution, or guidance if the point of a problem is not clear. In our experience, the Problems and Answers sections contain a great deal of the learning value of the book.

At Caltech we have used the book as a text for the final portion of a two-term Biochemistry course. It also should be suitable for courses in Molecular Biology, or as a supplementary text for broader courses in Genetics or Cell Biology.

Acknowledgements

Many people have contributed invaluable advice and assistance to the making of this book. We are indebted to our colleagues Eric Davidson, Richard Davidson, Margaret Dayhoff, Richard Dickerson, Sally Elgin, Walter Fitch, Joel Gottesfeld, Norman Horowitz, Joel Huberman, Ed Lewis, Richard McIntosh, Barbara Migeon, Donald Roufa, and John Stouffer for critically reading and commenting on individual chapters. We owe special thanks to Robert Benbow, Welcome Bender, Glen Galau, and Bill Kline for their assistance in writing and reviewing portions of the text. We are particularly grateful to Russell F. Doolittle for providing us with his article on protein evolution in advance of its publication; indeed, many features of Chapter 7 are based on that manuscript. We also thank the students and teaching assistants in the Caltech Biochemistry course for their patience and helpful suggestions in using preliminary editions of this book. We are grateful to Linda Wilson for converting our sketches for artwork into suitable form for final drawing, and to our many colleagues who gave us permission to use their illustrations. John Nelson has been an exceptional editor and valued advisor through all stages of the book's preparation. The book probably never would have been written without two summers of seclusion in the stimulating atmosphere of the Benjamin Writing Center at Aspen, Colorado, provided by Senior Editor Jim Hall and W. A. Benjamin, Inc. Finally, we deeply appreciate the support and tolerance of our families in coping with the often trying demands of authorship.

L. E. Hood
J. H. Wilson
W. B. Wood

CONTENTS

1 Genetic Organization And Control In Procaryotes

The details of genetic organization and control in procaryotes form a conceptual background for understanding how eucaryotes regulate their gene expression. This chapter considers the organization of procaryotic genomes, the operon model for transcriptional control, the molecular nature of control elements, and procaryotic models for development and differentiation.

ESSENTIAL CONCEPTS

1-1 Bacteria carry two kinds of genetic elements

(a) Most genes of a bacterium such as *E. coli* are carried on a single circular DNA molecule called the bacterial chromosome. Some strains of *E. coli* contain additional genes on smaller autonomously replicating circular DNA molecules called *plasmids* or *episomes* (Figure 1-1).

One such episome is the fertility factor, F. It carries genes that control bacterial mating or *conjugation*. Strains containing F are designated male (F⁺), and strains lacking F are designated female (F⁻). When F⁺ and F⁻ bacteria conjugate, the male synthesizes a copy of F and transfers it to the female via a cytoplasmic bridge formed between the two cells.

(b) By intracellular recombination, an F factor occasionally can incorporate a few genes from the bacterial chromosome. Such a modified fertility factor is designated F′ (Figure 1-1). During conjugation an F′ male cell transfers the incor-

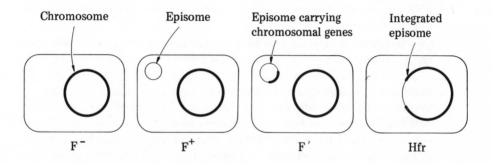

Chromosome	Episome	Episome carrying chromosomal genes	Integrated episome
F⁻	F⁺	F′	Hfr

Figure 1-1. Genetic elements and mating types of *E. coli*.

1

porated chromosomal genes to the female as part of the F′ factor. The recipient cell, now male, is known as a *merodiploid* (partial diploid), since it contains two copies of the transferred chromosomal genes: one copy on the recipient chromosome, and the second on the F′ episome. The merodiploid condition is relatively stable since the F′ factor, like the chromosome, replicates autonomously each time the cell divides.

(c) By intracellular recombination, an F factor occasionally can become incorporated into the bacterial chromosome. When such cells conjugate, the chromosome, now under the influence of the integrated F factor, is replicated and transferred into the female cell. Since strains with integrated F factors promote a large amount of genetic exchange, they are designated Hfr, for high-frequency recombination (Figure 1-1). Conjugating Hfr strains transfer the chromosome as a linear structure beginning at the point of integration of the F factor. Transfer of the entire chromosome requires 89 minutes at 37°C, and the integrated F factor is transferred last. If conjugating cells break apart before this time has elapsed, only a portion of the chromosome is transferred, and the recipient merodiploid remains F⁻. The transferred genes cannot replicate autonomously, and eventually are lost unless they become incorporated into the female chromosome by recombination. Hfr strains have been used to position a large number of genetic markers on the bacterial chromosome (Figure 1-2).

1-2 Bacterial structural genes can be nonregulated or regulated

(a) Genes that encode the structural information for specific proteins are called *structural genes*. Nonregulated structural genes code for enzymes and other proteins that the cell needs regardless of external conditions. Such products are synthesized *constitutively,* that is, at fixed relative rates. Constitutive synthesis requires only a single control element called a *promoter*. A promoter is a short stretch of DNA to which RNA polymerase must bind to initiate transcription of the adjacent structural gene. The relative rates of synthesis for various constitutive gene products are fixed by the relative affinities of their promoters for RNA polymerase. Affinity apparently is determined by the nucleotide sequence of the promoter.

(b) Regulated structural genes code for proteins that are required only under certain conditions. Their rates of synthesis are regulated according to the levels of metabolites in the external medium. In general, genes that code for degradative enzymes specific for certain fuel molecules are activated or *induced* by these molecules, whereas genes that code for biosynthetic enzymes are deactivated or *repressed* by the corresponding biosynthetic end products. For example, the sugar lactose in the medium induces synthesis of the hydrolytic enzyme β-galactosidase, whereas tryptophan (Trp) in the medium represses synthesis of the enzymes that catalyze Trp biosynthesis.

1-3 Regulated genes that are functionally related often occur in clusters

Groups of contiguous genes in *E. coli* control the enzymes of histidine (His) synthesis, Trp synthesis, arabinose utilization, and a number of other metabolic pathways (Figure 1-2). Functionally related gene clusters often are co-transcribed into a single polycistronic messenger RNA (mRNA) molecule that carries the information for all the proteins in the cluster. The unit of transcription in procaryotes is called an *operon*. Known operons range in size from one to twenty genes.

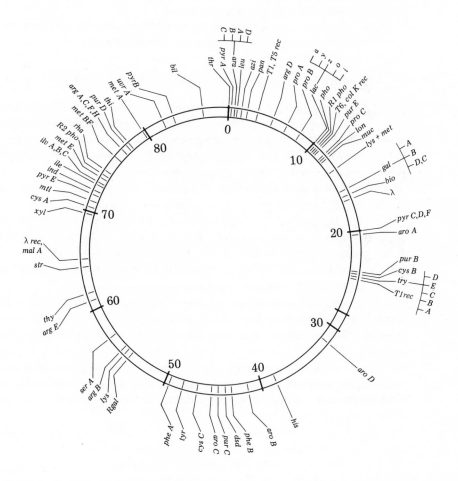

Figure 1-2. A simplified genetic map of *E. Coli*. Numbers on the inner circle indicate the time in minutes at which markers are transferred to an F⁻ cell by an Hfr strain that transfers *thr* as the initial marker. About 500 genes now have been mapped in *E. coli*. (Adapted from A. L. Taylor and M. S. Thoman, *Genetics* **50**, 667, 1964.)

1-4 Operon expression is controlled at the level of transcription

(a) The proteins of an operon are *coordinately controlled* by regulation of operon transcription. Since mRNA molecules in bacteria are degraded rapidly, rates of protein synthesis depend on the rates of corresponding mRNA synthesis.

(b) Transcriptional control of an operon requires at least three genetic regulatory elements:

    (1)   A *regulatory gene* that controls the synthesis of a regulatory protein.

    (2)   A *promoter* to which RNA polymerase must bind specifically to initiate transcription of the operon.

    (3)   An *operator,* a short stretch of DNA adjacent to the promoter to which the regulatory protein specifically binds.

1-5 The lac system in E. coli provided the initial model for operon control

(a) Jacques Monod and Francois Jacob originally formulated the operon model on the basis of experiments on lactose utilization, which is controlled by the *lac* operon. The *lac* operon includes three co-transcribed structural genes, designated *z, y,* and *a,* that code for the enzymes β-galactosidase, β-galactoside permease, and thiogalactoside transacetylase, respectively. β-galactosidase catalyzes hydrolysis of lactose into galactose and glucose; β-galactoside permease promotes the transport of β-galactosides into the cell; and thiogalactoside transacetylase performs a still unknown function that is not essential for lactose utilization. The regulatory gene controlling the operon is designated *i.* The genetic map order of these genes, the promoter *(p),* and the operator *(o)* is diagramed in Figure 1-3. The *lac* operon is *inducible;* it is transcribed only when a β-galactoside *(inducer),* such as lactose, is present in the medium.

(b) The nature of the *lac* operon regulatory elements was deduced from the effects of mutations in the operon. Normal (wild-type) cells are designated lactose positive or inducible *(lac⁺)*, since they can metabolize lactose when the operon is induced.

    (1)   Mutations in the *z* or *y* genes lead to a lactose negative *(lac⁻)* phenotype. Such mutant cells cannot metabolize lactose. Most mutations

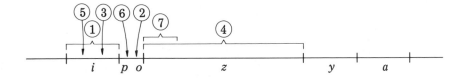

Figure 1-3. Map of the *lac* operon, showing locations of the mutations described in Table 1-2.

Table 1-1. Expression of the lactose operon in normal and mutant cells.

Phenotype	Synthesis of β-galactosidase and β-galactoside permease	
	in the absence of lactose	in the presence of lactose
lac^+ (normal, inducible)	–	+
lac^- (noninducible)	–	–
lac^c (constitutive)	+	+

in the i gene (i^-) produce a lactose constitutive (lac^c) phenotype, that is, a cell in which the operon is transcribed regardless of the presence or absence of inducer (Table 1-1). Rare mutations in the i gene (i^s) produce a lac^- phenotype. Mutations in the operator (o^c) also produce a constitutive phenotype. Mutations in the promoter (p^-) produce a lac^- phenotype.

(2) The distinction between i^- and o^c constitutive mutants and the roles of the various regulatory elements were deduced from experiments with merodiploids that included both a normal and a mutant form *(allele)* of the regulatory element. The phenotype of such a mero-diploid indicates which of two alleles is *dominant,* and which is *recessive.* As shown in Table 1-2, the mutant allele i^- is recessive to i^+, whereas the mutant o^c is dominant over o^+. In addition, distinctive *position effects* are observed. The i^+ allele exerts its effect on either a chromosomal or an episomal z gene; it therefore is said to be *trans*-dominant. By contrast, the o^c allele exerts its effect only on the z gene carried on the same genetic element; it therefore is said to be *cis*-dominant.

(3) These genetic experiments demonstrated that the *lac* operon is *negatively controlled* by its regulatory gene. That is, the product of the i gene is a repressor that prevents transcription of the operon in the absence of an inducer. In the presence of an inducer, the repressor does not function and transcription is permitted.

1-6 Four basic mechanisms of operon control are possible

(a) Many operons are controlled differently than the *lac* system. Operons in which the regulatory protein acts to *promote* transcription rather than prevent it are said to be *positively* rather than negatively controlled. Operons in which the controlling metabolite acts to *prevent* transcription rather than permit it are called *repressible,* rather than inducible. Thus there are four possible kinds of control systems involving a regulatory protein (P) and a controlling metabolite (M), as shown in Figure 1-4. So far, only three of the four kinds have been found in bacteria.

(b)   Examples of negatively controlled repressible operons are those controlling biosynthesis of His, Trp, and several other amino acids. The regulatory proteins of these operons prevent transcription, but only in the presence of the controlling metabolite. The controlling metabolite, called the *co-repressor,* is the amino acid itself or its aminoacyl-tRNA derivative. The molecular mechanisms of repression in these systems are not yet understood, partly because genetic identification of the regulatory proteins has proven to be more difficult than in the inducible systems. However, there is recent indirect evidence that in several operons controlling amino acid biosynthesis, the first enzyme in the pathway also may function as the repressor. Such systems are called *autoregulatory* since transcription is regulated by a product of the operon.

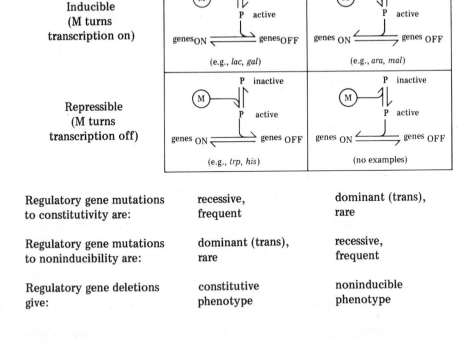

Figure 1-4.   Symmetry diagram of possible operon control systems involving a regulatory protein and a controlling metabolite. M represents a controlling metabolite and P represents a regulatory protein.

Table 1-2. Characteristics of *lac* operon mutants.

Class of mutation	Phenotype	Dominance	Frequency	Merodiploid phenotypes	
① i^-	lac^c	Recessive	Common	$i^+z^-/$ ① z^+	lac^+, inducible
				$i^-z^+/$ ① z^-	lac^+, inducible
② o^c	lac^c	Dominant (cis)	Rare	$i^+z^-/$ ② z^+	lac^c, constitutive
				$i^+z^+/$ ② z^-	lac^+, inducible
③ i^{-D}	lac^c	Dominant (trans)	Very rare	$i^+z^-/$ ③ z^+	lac^c, constitutive
				$i^+z^+/$ ③ z^-	lac^c, constitutive
④ z^-	lac^-	Recessive	Common	i^+z^+/i^+z^-	lac^+, inducible
				i^-z^+/i^-z^-	lac^c, constitutive
⑤ i^s	lac^-	Dominant (trans)	Very rare	$i^+z^+/$ ⑤ z^-	lac^-, noninducible
				$i^+z^-/$ ⑤ z^+	lac^-, noninducible
⑥ p^-	lac^-	Dominant (cis)	Very rare	$i^+z^+/$ ⑥ z^-	lac^+, inducible
				$i^+z^-/$ ⑥ z^+	lac^+, noninducible
⑦ z^- polar	lac^-	Dominant (cis)	Rare	$i^+z^+y^+/$ ⑦ y^-	lac^+, inducible
				$i^+z^+y^-/$ ⑦ y^+	lac^-, (permease) (noninducible)

(c) The arabinose *(ara)* operon is an example of a positively controlled inducible system. Figure 1-5 indicates the arrangement of the regulatory elements and the three structural genes, B, A, and D, controlling the enzymes of arabinose catabolism. Most mutations in the regulatory gene C produce a non-inducible *ara*⁻ recessive phenotype, as predicted if the product of the C gene is required as an activator to promote transcription of the operon. However, other evidence suggests that the C protein may act as a negative element as well, actively preventing transcription in the absence of the inducer arabinose.

1-7 Catabolite repression provides for override control of certain operons

Bacteria employ a general positive control system, known as *catabolite repression,* that overrides the transcriptional control of many operons related to sugar metabolism. In the presence of glucose, induction of enzymes involved in the breakdown of other sugars is blocked, even if these sugars are present in amounts that normally would cause induction. The controlling metabolite for the override system is 3', 5'-cyclic AMP (cAMP), which decreases in concentration when glucose is present and increases when it is absent. The action of cAMP is mediated through a catabolite activator protein (CAP), which in the presence of cAMP binds to the promoters of operons such as *lac* and *ara*. Transcription of these operons can occur only when this protein is bound, regardless of the presence or absence of the operon-specific regulatory proteins. Glucose, by decreasing the cAMP level, inactivates the activator protein, thereby preventing transcription of the *lac* and *ara* operons.

1-8 The molecular nature of the lac operon control elements is known

(a) The repressor of the *lac* operon, controlled by the *i* gene, has been purified and characterized. It is a tetrameric protein of identical 40,000-molecular-weight subunits, each of which carries a single binding site for inducer. The natural inducer, allolactose, binds noncooperatively with a dissociation constant (K_D) of ~$10^{-6}M$ and causes a conformational change in the repressor molecule. In the absence of inducer, the repressor binds specifically to DNA carrying the *lac* operator $(K_D \simeq 10^{-12}M)$; the half-time for dissociation of this complex is about 20 minutes. In the presence of inducer, both the K_D and the dissociation rate of the repressor-operator complex are increased by several orders of magnitude. Low-resolution x-ray diffraction studies show that the repressor molecule is quite elongated and suggest that DNA complexes with it like a hot dog in a bun. This interpretation is consistent with the length of the DNA sequence to which the repressor binds (Figure 1-6).

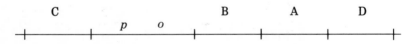

Figure 1-5. Order of genes in the arabinose operon.

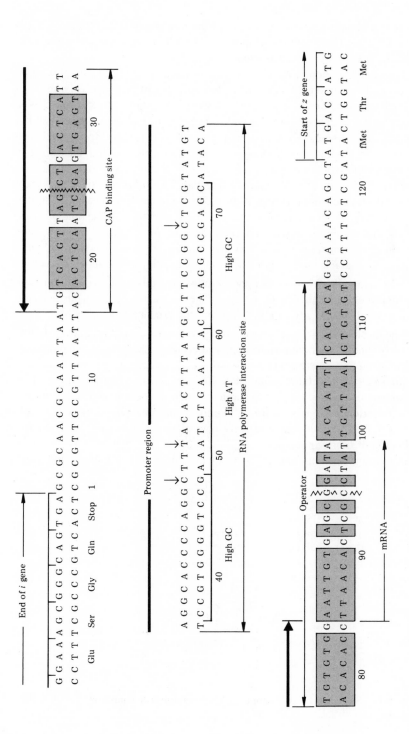

Figure 1-6 Nucleotide sequence of *lac* operon control elements. Wavy lines indicate two-fold rotational symmetry axes. Screened nucleotides are arranged palindromically (Figure 1-7) about an axis of symmetry. Vertical arrows indicate the sites of nucleotide substitution in three known promoter mutations. Exact boundaries for the various binding sites are not known and have been drawn somewhat arbitrarily. (Adapted from R. C. Dickson *et al.*, *Science*, **187**, 27, 1975.)

(b)   The catabolite activator protein is a dimer of identical 23,000-molecular-weight subunits. The details of its binding to DNA in the presence and absence of cAMP are not yet well characterized.

(c)   The control region of the *lac* operon is a sequence of 122 nucleotide pairs that extends from the end of the *i* gene to the beginning of the *z* gene (Figure 1-6). The binding sites for *lac* repressor, catabolite activator protein, and RNA polymerase have been shown to lie within this region by genetic studies and by sequence analysis of DNA that is protected specifically from nuclease degradation by binding of *lac* repressor and RNA polymerase. Both the activator protein and repressor binding sites have two-fold rotational axes of near-perfect symmetry. Such symmetrical sequences are called *palindromes,* since they read the same backwards and forwards if the reader switches from one strand to the other at the axis of symmetry (Figure 1-7). The symmetry of these binding sites suggests that catabolite activator protein and the *lac* repressor possess correspondingly symmetrical recognition sites. Such corresponding symmetry may be a general characteristic of DNA binding by multimeric proteins.

(d)   Regulatory elements in other operons are less well characterized. Only the activator protein controlled by the C gene of the *ara* operon has been purified and shown to bind specifically to DNA carrying the *ara* operator.

(e)   The mechanisms by which regulatory proteins act are not yet known. In the negatively controlled operons so far analyzed in detail, the operator and promoter sequences overlap or are immediately adjacent, suggesting that the repressor may physically block the binding of RNA polymerase to the promoter. The mechanisms of positive regulatory proteins such as the *ara* activator and the catabolite activator are more difficult to understand. One possibility with some experimental support is that positively controlled operons carry a signal for termination of transcription adjacent to the promoter, so that RNA polymerase alone can transcribe only a short DNA sequence near the beginning of the oper-

(a)   M A D A M   I' M   A D A M

(b)   A T A G   C T A T

T A T C   G A T A

Figure 1-7. Structure of palindromes. (a) An English palindrome. (b) A DNA palindrome.

on. The positive regulatory protein then acts as an *antiterminator,* which some-how prevents termination and thus allows transcription of the entire operon.

1-9 Irreversible sequences of procaryotic gene expression provide models for eucaryotic development

Development of a eucaryotic embryo involves readout of a genetic program, that is, an irreversible sequence of gene expression that continues to completion once it is set in motion. Analogous although much simpler temporal regulation is seen in two kinds of procaryotic systems: bacteriophage infection and bacterial sporu-lation. Both systems have been studied extensively in the hope of providing insights into the more complex control of eucaryotic development. Several dif-ferent mechanisms for temporal control of gene expression have been observed in procaryotes.

(1) Even the simplest bacterial viruses, the RNA bacteriophages (WWBH Chapter 17), show some temporal control of their gene ex-pression. Whereas production of the major and minor coat proteins begins immediately after infection, synthesis of RNA synthetase is delayed for about four minutes, then proceeds linearly from the fourth to the tenth minute, and then ceases, while synthesis of the other two proteins continues. This control must be at the translational level, since no transcription is involved. The four-minute delay is due to the necessity of translating the coat-protein gene in order to open up the secondary structure of the RNA sufficiently to allow ri-bosome binding to the synthetase initiation site. The shutoff of syn-thetase synthesis is accomplished by specific binding of newly syn-thesized coat-protein subunits to the RNA at a site near the start of the synthetase gene, thereby blocking its translation after 10 minutes.

(2) The simple single-stranded-DNA phages such as ϕX174 appear to be the only systems with no temporal control of gene expression. The entire genome (nine genes) is transcribed into a single mRNA mole-cule, and the phage proteins all appear at about the same time. There are timed events in the infection (for example, double-stranded-DNA synthesis stops and single-stranded-DNA synthesis begins at approximately 12 minutes after infection), but this is probably due to interaction of coat-protein subunits with replicating DNA, rather than to gene regulation.

(3) T7, a small, double-stranded DNA bacteriophage, expresses its four "early" genes immediately following infection, then shuts these off and expresses the remaining 20 "late" genes, including those for the structural proteins of the virus. This regulation is accomplished by a simple positive control mechanism. Following infection, the bacte-rial RNA polymerase transcribes the four early genes. One of these genes codes for a new RNA polymerase, a monomeric enzyme of

about 110,000 molecular weight, which specifically recognizes promoters for the "late" genes. Concomitant inactivation of the bacterial enzyme by an unknown mechanism shuts off transcription of the early genes.

(4) Lambda, a somewhat larger DNA phage with about 50 genes, has two alternative fates following infection. Its DNA either can incorporate itself into the bacterial chromosome and remain dormant (lysogenization), or it can replicate, leading to production of more virus and lysis of the host cell (lytic infection). Only lytic infection is considered here. It is under the control of at least two positive elements, the products of genes N and Q (Figure 1-8). The N gene is transcribed by the bacterial RNA polymerase immediately after the onset of lytic infection. The resulting N protein promotes transcription of the early gene operon, at the distal end of which is gene Q. The

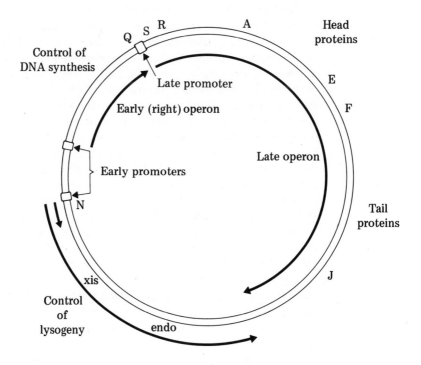

Figure 1-8. Operons of bacteriophage lambda. (Adapted from James D. Watson, *Molecular Biology of the Gene,* 2nd ed., © 1970 by J. D. Watson, W. A. Benjamin, Inc.)

resulting Q protein then promotes transcription of the late gene oper-on, which includes the genes for phage structural proteins. The mechanisms by which the N and Q regulatory proteins act are not yet certain, but there is some evidence that they function as antiterminators, as described in Essential Concept 1-8(e).

(5)   T4, a large bacteriophage with about 150 genes, shows more com-plex control. It has a number of operons, and at least four classes of genes, termed immediate-early, delayed-early, quasi-late, and late, according to their time of expression after infection. Immediate- and delayed-early genes appear to be located in the same early operons. The time differences in appearance of the corresponding proteins can be explained by the relative positions of genes in these operons; immediate-early genes are promoter-proximal, whereas delayed-early genes are promoter-distal. Expression of the quasi-late genes requires recognition of a new set of promoters. Later in infection, a third set of promoters is recognized, thereby resulting in transcript-ion of late operons, while transcription of quasi-late genes decreases.

These changes in gene expression are accompanied by alter-ations in the bacterial RNA polymerase (see WWBH Essential Concept 18-2). Shortly after infection, the α subunit of the host en-zyme becomes adenylated, and the core enzyme loses its affinity for the σ subunit. Later in infection, the β subunits are altered, probably by limited proteolysis. At about the same time, the product of the phage gene 55 (P55), whose presence is required for late gene ex-pression, becomes associated with the core polymerase.

The relationships between these alterations are not yet estab-lished. However, it appears possible that temporal regulation of gene expression is accomplished by successive alterations of the RNA polymerase core complex, with concomitant replacement of positive control elements (e.g., σ, P55) that allow recognition of successive classes of promoters.

(6)   Similarly controlled sequences of gene expression occur in spore-forming bacteria. An appropriate change in the environment some-how turns off the genes that maintain vegetative growth and turns on the genes that control sporulation. In *Bacillus subtilis,* the bacterial species most extensively studied, the initial events appear analogous to those in T4 infection. The RNA polymerase becomes altered so that it no longer will bind the σ factor of the vegetative cell, and consequently no longer will recognize promoters of vegetative oper-ons. A new RNA polymerase subunit, analogous to P55 of T4, may be synthesized to allow recognition of new promoters, but the evidence for this mechanism is not yet conclusive.

1-10 Additional concepts and techniques are presented in the Problems section

(a) Molecular explanations for some rare *lac* regulatory mutants. Problem 1-4.

(b) Natural inducer for the *lac* operon. Problem 1-5.

(c) Purification of the *lac* repressor. Problem 1-14.

REFERENCES

Where to begin

G. S. Stent, *Molecular Genetics, An Introductory Narrative* Chapters 10, 11, 19, and 20 (Freeman, San Francisco, 1971)

J. D. Watson, *Molecular Biology of the Gene* Chapters 14 and 15 (W. A. Benjamin, Menlo Park, Calif., 1970, 2nd ed.)

Bacterial genetics and chromosome regulation

W. F. Hayes, *The Genetics of Bacteria and their Viruses* (Wiley, New York, 1968, 2nd ed.)

Bacteriological Reviews **36** #4 (December, 1972). (Articles by A. L. Taylor and C. D. Trotter; K. B. Low; and K. E. Sanderson *et al.*)

Chromosome Structure and Function: Cold Spring Harbor Symposia on Quantitative Biology, Vol. 38, pp. 31-58 (Cold Spring Harbor Laboratory, Cold Spring Harbor, N. Y., 1973)

Gene regulation

J. R. Beckwith and D. Zipser (Ed.), *The Lactose Operon* (Cold Spring Harbor Laboratory, Cold Spring Harbor, N.Y., 1970)

R. F. Goldberger, "Autogenous regulation of gene expression," *Science* **183,** 810 (1974)

F. Jacob and J. Monod, "Genetic regulatory models in the synthesis of proteins," *J. Mol. Biol.* **3,** 318 (1961)

W. S. Reznikoff, "The operon revisited," *Ann. Rev. Genetics* **6,** 133 (1972)

Transcription of Genetic Material: Cold Spring Harbor Symposia on Quantitative Biology, Vol. 35 (Cold Spring Harbor Laboratory, Cold Spring Harbor, N. Y., 1970)

Molecular nature of genetic control elements

M. J. Chamberlin, "The selectivity of transcription," *Ann. Rev. Biochem.* **43,** 721 (1974)

R. C. Dickson, J. Abelson, W. M. Barnes, and W. S. Reznikoff, "Genetic regulation: the *lac* control region," *Science* **187,** 27 (1975)

W. Gilbert, N. Maizels, and A. Maxam, "Sequences of controlling regions of the lactose operon," *Cold Spring Harbor Symposium on Quantitative Biology* **38,** 845 (1973)

T. Maniatis, M. Ptashne, and R. Maurer, "Control elements in the DNA of bacteriophage λ," *Cold Spring Harbor Symposia on Quantitative Biology* **38,** 857 (1973)

Temporal control of sequential gene expression

R. Calendar, "The regulation of phage development," *Ann. Rev. Microbiol.* **24,** 241 (1970)

H. Echols, "Developmental pathways for the temperate phage: lysis vs. lysogeny," *Ann. Rev. Genetics* **6,** 157 (1972)

A. D. Hershey (Ed.), *The Bacteriophage* λ (Cold Spring Harbor Laboratory, Cold Spring Harbor, N. Y., 1971)

I. Herskowitz, "Control of gene expression in bacteriophage lambda," *Ann. Rev. Genetics* **7,** 289 (1973)

R. Losick, "*In vitro* transcription," *Ann. Rev. Biochem.* **41,** 409 (1972)

P. Z. O'Farrell and L. M. Gold, "Bacteriophage T4 gene expression," *J. Biol. Chem.* **248,** 5502 (1973)

W. C. Summers, "Regulation of RNA metabolism of T7 and related phages," *Ann. Rev. Genetics* **6,** 191 (1972)

PROBLEMS

1-1 Answer the following with true or false. If false explain why.

(a) Bacteria that carry the fertility factor F are designated female.

(b) Bacteria that carry F act as donors of genetic material in bacterial conjugation.

(c) When an F′ bacterium and an F⁻ bacterium conjugate, the episome is transferred from one to the other, thereby transforming the donor into a female cell and the recipient into a male.

(d) Bacteria that carry an F′ episome are usually diploid for some chromosomal genes.

(e) Stable merodiploids are almost always female cells.

(f) Most mutations in the i gene of the *lac* operon prevent the utilization of lactose as a carbon source.

(g) In merodiploids, i-gene mutations are always recessive, whereas operator mutations are always dominant.

(h) The *lac* operon is under negative control by its regulatory gene i.

(i) β-galactosides induce the *lac* operon by combining with the repressor and promoting its binding to the operator.

(j) If you discovered a positively controlled repressible operon, you would expect that most mutations in the regulatory gene would lead to expression of the structural genes even in the presence of co-repressor.

(k) Two unlinked bacterial operons can be under the control of the same regulatory gene.

(l) In the presence of arabinose plus glucose, transcription of the *ara* operon will occur only if cAMP is added to promote binding of the catabolite activator protein to the *ara* promoter.

(m) In most negatively controlled operons, the promoter lies between the operator and the first structural gene.

(n) If a bacterium is infected with two mutant λ bacteriophages, one defective in gene N and the other in gene Q, neither mutant will reproduce.

(o) Hairpin structures in RNA indicate the presence of palindromic sequences in the DNA from which the RNA was transcribed.

1-2 (a) The fertility factor, F, replicates autonomously as an _____ except in Hfr strains.

(b) An F factor carrying chromosomal genes is symbolized by _____ .

(c) A cell that has received an F′ episome by conjugation is called a _____ because it contains duplicate copies of the chromosomal genes carried on the episome.

(d) Operons can be defined as units of _____ .

(e) To initiate transcription of an operon, RNA polymerase must bind to a site on the DNA called the _____ .

(f) In operons under _____ control, binding of the regulatory protein to the operator may promote RNA polymerase binding to the promoter, thereby allowing transcription to occur.

(g) An operon is said to be _____ if the presence of a specific metabolite prevents transcription of the operon.

(h) Expression of a gene is said to be _____ when it is not affected by the presence or absence of metabolites in the external medium.

(i) Mutations in the regulatory gene that lead to constitutive transcription are rare in operons under _____ control.

(j) Lactose cannot induce transcription of the *lac* operon unless sufficient _____ is present to combine with _____ , thereby allowing its binding to the *lac* promoter.

(k) Sequential control of gene expression in procaryotes seems to be mediated primarily by _____ control elements.

1-3 What will be the phenotypes (*lac*$^+$, inducible; *lac*$^-$, noninducible; or *lac*^{c}, constitutive) of merodiploids with the following configurations of *lac* alleles? (Assume that all genes not indicated are present as the + allele.)

(a) $i^s p^+ z^- / i^+ p^- z^+$

(b) $i^- z^- / i^s z^+$

(c) $i^s o^+ z^- / i^+ o^c z^+$

(d) $p^- o^+ z^+ / p^+ o^c z^-$

(e) $p^- o^+ z^- / p^+ o^c z^+$

1-4 Explain in molecular terms how point mutations in the *i* gene of the *lac* operon can lead to the following phenotypes.

(a) i^-: β-galactosidase is made constitutively, but this character is recessive to the wild-type allele; merodiploids of genotype i^+z^+/i^-z^+ make enzyme only when inducer is present, like the wild-type i^+z^+ strain.

(b) i^s: enzyme is not made either in the presence or absence of inducer, and this character is *trans*-dominant; i^+z^+/i^sz^+ merodiploids are noninducible.

(c) i^{-D}: enzyme is made constitutively as in (a), but the character is dominant; the corresponding merodiploid also makes enzyme constitutively.

1-5   Allolactose is an isomer of lactose that is produced by β-galactosidase as an intermediate in the cleavage of lactose to galactose and glucose (Figure 1-9). It is allolactose, not lactose, that is the natural inducer for the *lac* operon. Allolactose binds to the repressor and lowers its affinity for the *lac* operator, thereby opening the operon for transcription.

(a)   Would you expect β-galactoside permease to be induced in a z^-y^+ mutant upon addition of lactose? Upon addition of allolactose?

(b)   Would you expect β-galactosidase to be induced in a z^+y^- mutant upon addition of lactose? Upon addition of allolactose?

(c)   Can you suggest a reason why cells synthesize low levels of β-galactosidase and β-galactoside permease even when there is absolutely no lactose in the medium?

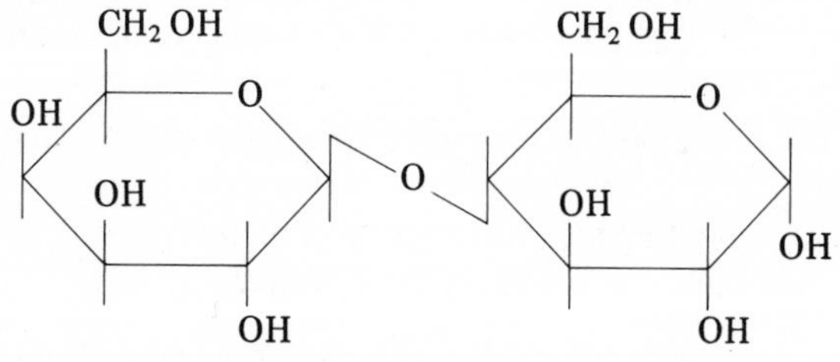

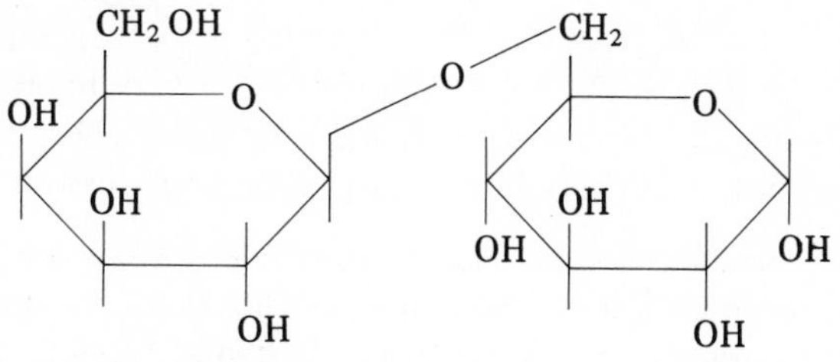

Figure 1-9.   Structures of lactose and allolactose (Problem 1-5).

1-6 The Trp operon in *E. coli* is under the control of a regulatory gene, *trp*R.

(a) Complete Table 1-3 to indicate the phenotype of wild-type (R^+) cells, and the mutant phenotype you would expect to result from the majority of mutations in the R gene (R^- mutations).

(b) Describe the phenotype of the merodiploid R^-/R^+.

Table 1-3. Phenotypic consequences of regulatory-gene mutations in the *trp* operon (Problem 1-6).

Cells	Synthesis of Trp enzymes		Phenotype
	in the absence of Trp	in the presence of Trp	
R^+			
R^-			

1-7 Suppose you discovered a mutant carrying an alteration in the regulatory gene (R) for the Trp operon that prevented operon expression in either the presence or absence of Trp. Call the mutant allele R^x, and assume that the R gene codes for a repressor protein.

(a) What is the most likely molecular explanation for the phenotype of an R^x mutant?

(b) What would be the phenotype of a merodiploid R^x/R^+, carrying both the mutant and wild-type alleles of the R gene?

1-8 An enzyme required for the synthesis of proline (Pro) in a soil bacterium normally is made only when Pro is absent from the growth medium (Table 1-4). When mutants with defects in control of this enzyme are isolated, two classes of mutants altered in a regulatory gene R are found with the frequencies and phenotypes shown in Table 1-4. S is the structural gene for the enzyme.

(a) From the results in Table 1-4, what is the most likely nature of the control system (inducible, negative; repressible, negative; inducible, positive; or repressible, positive)? Explain your reasoning very briefly.

(b) Complete Table 1-4 to show the most likely phenotypes of the indicated merodiploids, again very briefly explaining your answers.

1-9 What will be the phenotypes (*ara*$^+$, inducible; *ara*$^-$, noninducible; or *ara*^{c}, constitutive) of F$^-$ cells and F$'$ merodiploids with the following configurations of *ara* alleles? (Assume that all genes not indicated are present as the + allele.)

(a) C^+B^+

(b) C^-B^+

(c) C^+B^-

(d) $C^+o^-B^+$

Table 1-4. Synthesis of a Pro biosynthetic enzyme in normal and mutant bacteria (Problem 1-8).

Strain	Mutant frequency	Enzyme synthesis Pro present	Pro absent
R^+S^+ (Normal)	–	–	+
R^+S^-	10^{-5}/cell/generation	–	–
R^-S^+	10^{-7}/cell/generation	–	–
R^cS^+	10^{-5}/cell/generation	+	+
R^+S^-/R^-S^+	–		
R^cS^+/R^+S^-	–		
R^cS^+/R^-S^-	–		

(e) $C^+p^-B^+$
(f) C^+B^-/C^-B^+
(g) C^+B^+/C^-B^-
(h) $C^+o^-B^+/C^+o^+B^-$
(i) $C^+p^-B^+/C^+p^+B^-$
(j) $C^+o^-B^-/C^-o^+B^+$

1-10   (a)   Rationalize the existence of the catabolite repression override system in terms of its usefulness to the cell.

(b)   Predict the phenotypic characteristics that would result from two different kinds of mutation in the gene for the catabolite activator protein, indicating which of these kinds you would expect to be more frequent.

1-11   At the molecular level, does catabolite repression represent a positive inducible or positive repressible control system? Explain.

1-12   Consider a regulated bacterial system consisting of a regulatory gene (r), a promoter (p), an operator (o), and a structural gene (s), carrying information for an easily measurable enzyme activity. For each of the following questions, indicate whether the enzyme will be constitutively produced, inducible, repressible, or absent under all conditions, in both the mutant indicated and a merodiploid made by introducing the normal regulatory elements (i.e., an episome of genotype $r^+p^+o^+s^-$) into the mutant.

(a)   If the system is *positively* controlled by the product of gene r and inducible by some metabolite, what are the ways in which enzyme production can be affected by

(1)   deletion mutations in r.
(2)   deletion of p.

Table 1-5. Expression of X synthetase in wild-type and mutant cells (Problem 1-13).

Cell	X synthetase production	
	in absence of X	in presence of X
Wild-type	+	−
Mutants ①, ②, ③, ④, ⑥, ⑦, ⑧, ⑨, ⑩	−	−
Mutant ⑤	+	+

(b)   If the system is *negatively* controlled and repressible, what are the ways in which enzyme production can be affected by
   (1)   deletion mutations in *r*.
   (2)   point mutations in *o*.
   (3)   point mutations in *r*.

1-13   You are studying control of natural product, designated simply compound X, in a soil bacterium. You know that the enzyme X synthetase is coded by a single structural gene, designated *s*, and that the expression of this gene is affected by the presence or absence of X in the medium, as shown in Table 1-5. To investigate control of the synthetase, you isolate several mutants that show altered patterns of enzyme production. The phenotypes of these mutants also are shown in Table 1-5. You then map the corresponding mutations, and find their positions relative to the *s* gene as shown in Figure 1-10. Finally, you construct the merodiploids shown in Table 1-6, using strains carrying one or two mutant alleles on their chromosome, and determine the resulting phenotypes.
   (a)   Is the X operon inducible or repressible by X?
   (b)   Is the operon under positive or negative control?
   (c)   Which mutations probably alter the regulatory protein?
   (d)   Which mutation might be a defect in the operator?
   (e)   Which mutation might be a defect in the promoter?
   (f)   Briefly describe in molecular terms how the operon is controlled in the wild-type, and give a molecular explanation for the effects of mutations 1, 5, and 9.

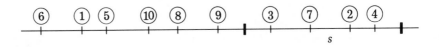

Figure 1-10.   Map positions of mutations carried by mutants described in Table 1-5 (Problem 1-13).

Table 1-6. Expression of X synthetase in merodiploids (Problem 1-13).

| | X synthetase production | |
Merodiploid	in absence of X	in presence of X
++/++ (wild-type alleles)	+	–
⑦ / ①	+	–
++/ ① ⑦	+	–
⑦ / ⑤	+	+
++/ ⑤ ⑦	+	+
⑦ / ⑥	+	–
++/ ⑥ ⑦	+	–
⑦ / ⑧	+	–
++/ ⑧ ⑦	+	–
⑦ / ⑨	–	–
++/ ⑨ ⑦	+	–
⑦ / ⑩	+	–
++/ ⑩ ⑦	+	–

1-14   Positive and negative control systems require the same number of control elements, exhibit approximately the same degree of complexity, and are equally effective in regulation. Thus there seems to be no compelling reason to expect a particular operon to exhibit one or the other kind of control. However, there are no known examples of biosynthetic-enzyme operons that are under positive control. Can you rationalize this apparent selection against positive control on the basis of your understanding of mutational consequences in negatively and positively controlled operons?

1-15   For many years attempts to purify the *lac* repressor were stalled because of the lack of a suitable assay method. The approach taken assumed that the repressor must be able to bind β-galactoside inducers, such as isopropyl thiogalactoside (IPTG). Attempts were made to detect an IPTG-binding protein in *E. coli* extracts by equilibrium dialysis, as follows. A sample of extract was placed inside a dialysis sac and suspended in a reservoir of buffer containing labeled IPTG (which can pass freely through a dialysis membrane). When the system had reached equilibrium, aliquots of the external solution and the solution inside the sac were assayed for radioactivity. Since the concentration of *free* IPTG inside and outside the sac must be equal at equilibrium, a higher concentration of IPTG inside would indicate that some was bound to a large molecule in the extract.

Determine the feasibility of such an assay by the following calculation. Assume that the dissociation constant for the complex (RI) of repressor (R) and inducer (I) is

$$K_D = \frac{[R]\,[I]}{[RI]} = 10^{-6}M$$

and that [I] in the external solution at equilibrium is $10^{-6}M$, corresponding to a radioactivity of 1000 counts per minute (cpm)/ml. Given that 1 ml of extract corresponds to the contents of about 6×10^{11} cells, and that, as we now know, there are about 10 molecules of repressor per cell, what will be the level of cpm/ml inside the dialysis sac?

1-16   Find the longest three overlapping palindromes in the following sequence for the early (left) operator-promoter region in λ. The nucleotides indicated in parentheses are known to be a pyrimidine-purine pair, but their identity is not established. (From T. Maniatis *et al., Nature,* **250,** 394, 1974.)

```
        30          25          20          15          10          5           1
  C A C G A G T C A T A G T G G C G G T C A C C A T A A A T A C A G (C) T

  G T G C T C A G T A T C A C C G C C A G T G G T A T T T A T G T C (G) A
```

1-17   Dr. Ingrid Irgnird has been fascinated with palindromes ever since she received her M.D. degree. The discovery that palindromic DNA sequences may be important as specific protein binding sites piqued her curiosity in molecular biology. However, she was skeptical when she learned that only imperfect palindromes had been found. As a cautionary note against assuming that all palindromes have significance as protein binding sites, she gives you the following problem.

The following partially palindromic DNA sequence begins two nucleotide pairs to the right of the axis of symmetry of the *lac* repressor binding site:

```
A T A  A C  A A  T T T  C A C  A  C A G  G A A A  C A G  C  T A T

T A T  T G T  T  A A A G  T G T  G T C  C T T T  G T  C  G  A T A
```

(a)   What is the average frequency with which this same partial palindrome will occur by chance in a random DNA sequence the length of the *E. coli* genome (4 × 10⁶ nucleotide pairs)? Neglect the nucleotide pairs that are not palindromically arranged.

(b)   How many palindromes of this general character (16 of 28 nucleotide pairs arranged palindromically) will occur on the average in a random DNA sequence of 4 × 10⁶ nucleotide pairs?

(c)   If you set a frequency limit of 0.1 occurrences per random sequence of 4×10^6 nucleotide pairs, on the average, as discriminating between a randomly occurring palindrome and one that may be a protein binding site, how many of the nucleotide pairs in a 28-nucleotide-pair sequence would have to be arranged palindromically before you would consider the palindrome to be a probable binding site?

ANSWERS

1-1   (a)   False. Bacteria carrying the sex factor F are designated male.
(b)   True
(c)   False. The episome is replicated in the course of conjugation and only the copy is transferred into the female cell. Thus both donor and recipient carry F′ and are male following conjugation.
(d)   True
(e)   False. Stable merodiploids are almost always F′ males.
(f)   False. Most mutations in the i gene lead to constitutivity; the cell remains lac^+ but the lac enzymes are synthesized regardless of the presence or absence of an inducer.
(g)   False. At least two known i-gene mutations are dominant over the wild-type (see Table 1-2).
(h)   True
(i)   False. β-galactosides induce the lac operon by combining with the repressor, thereby preventing its binding to the operator.
(j)   False. Most mutations in the regulatory gene would lead to inability to express the structural genes even in the absence of co-repressor.
(k)   True
(l)   True
(m)   False. In the negatively controlled operons so far analyzed in detail, the operator either overlaps the promoter or lies between the promoter and the first structural gene.
(n)   False. Both genes N and Q code for positively acting regulatory proteins; therefore the two mutants will complement each other and both will be able to reproduce.
(o)   True

1-2   (a)   episome (or plasmid)
(b)   F′
(c)   merodiploid
(d)   transcription
(e)   promoter
(f)   positive
(g)   repressible

(h) constitutive
(i) positive
(j) cAMP, catabolite activator protein
(k) positive

1-3 (a) *lac⁻*, noninducible
(b) *lac⁻*, noninducible
(c) *lacᶜ*, constitutive
(d) *lac⁻*, noninducible
(e) *lacᶜ*, constitutive

1-4 (a) The i^- phenotype will be caused by any mutation in the i gene that leads to production of an inactive repressor protein, that is, a repressor that cannot bind to the operator.
(b) The i^s phenotype will be produced by mutations that prevent the binding of β-galactosides to the repressor protein without affecting its ability to bind to the operator. In merodiploid cells carrying the wild-type i^+ allele as well, the *lac* operon will be noninducible because the mutant protein will continue to bind to the *lac* operator whether or not β-galactosides are present.
(c) The i^{-D} phenotype is difficult to understand. It is explained best on the assumption that all the subunits of the repressor protein, which is a tetramer, must be functional if the repressor molecule is to bind to the operator. The i^{-D} mutation is thought to produce a repressor subunit that is nonfunctional in binding but still can participate in tetramer formation. The character is dominant because in cells containing both the i^{-D} and a wild-type allele, assuming random subunit association, the probability is high that most of the tetramers formed will have at least one defective subunit.

1-5 (a) Addition of lactose will not induce β-galactoside permease because lactose cannot be converted to allolactose in a cell that is defective in β-galactosidase. Addition of allolactose, however, will induce β-galactoside permease since it does not require the presence of β-galactosidase.
(b) Neither addition of lactose nor allolactose will induce β-galactosidase in a cell that is missing β-galactoside permease, because they cannot be transported into the cell in the absence of the transport protein.
(c) It is apparent from the foregoing examples that small amounts of β-galactosidase and β-galactoside permease must be present in the cell for induction to occur when lactose appears in the medium.

1-6 (a) See Table 1-7.
(b) The merodiploid phenotype will be *trp⁺*, repressible. Since the system is under negative control, like the *lac* system, R⁻ mutations leading to defective repressor will be recessive to the R⁺ wild-type allele.

1-7 (a) The mutation most likely leads to an altered repressor that is "stuck" in the active configuration and thus will bind to the operator even in the absence of Trp. This is somewhat analogous to the i^s mutation in the *lac* operon.

Table 1-7. Phenotypic consequences of regulatory-gene mutations in the *trp* operon (Answer 1-6).

Cells	Synthesis of Trp enzymes		Phenotype
	in the absence of Trp	in the presence of Trp	
R^+	+	−	*trp*$^+$, repressible
R^-	+	+	*trp*$^+$, constitutive

(b)   The merodiploid will be *trp*$^-$, since the mutant repressor will prevent operon transcription even if the normal repressor is inactive due to the absence of Trp.

1-8   (a)   Enzyme synthesis normally is *repressed* by Pro. Since the more common type of mutation in the R gene leads to constitutive enzyme synthesis, R is most likely a repressor and therefore control is *repressible, negative.*

(b)   If control is repressible, negative, the R^c mutation will be recessive to R^+ and R^-; R^- will be *trans*-dominant over R^+ and R^c. Hence the phenotypes are as shown in Table 1-8.

Table 1-8. Synthesis of a Pro enzyme in three merodiploids (Answer 1-8).

Strain	Enzyme synthesis	
	Pro present	Pro absent
R^+S^-/R^-S^+	−	−
R^cS^+/R^+S^-	−	+
R^cS^+/R^-S^-	−	−

1-9   (a)   *ara*$^+$, inducible
(b)   *ara*$^-$, noninducible
(c)   *ara*$^-$, noninducible
(d)   *ara*$^-$, noninducible
(e)   *ara*$^-$, noninducible
(f)   *ara*$^+$, inducible
(g)   *ara*$^+$, inducible
(h)   *ara*$^-$, noninducible
(i)   *ara*$^-$, noninducible
(j)   *ara*$^+$, inducible

1-10   (a)   Presumably, catabolite repression is useful to the cell in preventing the wasteful elaboration of enzymes for conversion of other sugars to glucose or products of glucose catabolism when glucose is already present.

(b)   Since the catabolite activator protein is a positive regulatory element, the majority of mutations in the corresponding gene should result in an inactive protein and a resulting inability to induce any of the sugar-metabolizing enzymes under catabolite repression control. A much rarer mutational alteration could lead to a catabolite activator that would be active even in the absence of the controlling metabolite cAMP. Such a mutation would produce a cell in which catabolite repression could not function, that is, in which glucose would have no inhibiting effect on induction of the sugar-metabolizing enzymes.

1-11   Catabolite repression is an example of a positive, inducible control system. Glucose acts physiologically as a co-repressor, but it does so indirectly by influencing the intracellular concentration of cAMP, which is the true controlling metabolite. cAMP acts as an inducer; it binds to the catabolite activator to form an active complex that binds to DNA and turns on transcription (provided that the operon-specific regulatory proteins are in the proper configuration).

1-12   (a)   (1)   Enzyme production will be noninducible in the mutant and inducible in the merodiploid.

               (2)   Enzyme production will be noninducible in both the mutant and the merodiploid.

      (b)   (1)   Enzyme will be produced constitutively in the mutant, and repressible in the merodiploid.

               (2)   Enzyme will be produced constitutively in both the mutant and the merodiploid.

               (3)   Enzyme production can be either the same as in (1), or, less frequently, absent under all conditions.

1-13   (a)   The operon is repressible by X (see wild-type, Table 1-5).

(b)   Since most mutations in the regulatory gene prevent X synthetase production even in the absence of X, and are recessive to the wild-type regulatory gene in a merodiploid, the operon must be under positive control.

(c)   Mutations 1, 5, 6, 8, and 10 probably alter the regulatory protein. Mutations 1, 6, 8, and 10 prevent X synthetase production under all conditions, but are recessive to the wild-type allele. Mutation 5 shows a different pattern but maps between 1 and 10 (Figure 1-10). Mutation 9 cannot be a regulatory gene mutation, since it is *cis*-dominant in a merodiploid (Table 1-6).

(d)   Mutation 9 could alter the operator so that it no longer is recognized by the positive regulatory protein.

(e)   Mutation 9 also could be a promoter defect that prevents RNA polymerase binding.

(f)   In the wild-type, when X is absent, the protein coded by the regulatory gene presumably binds to an operator adjacent to the structural gene, thereby allowing RNA polymerase to bind at the promoter site and to initiate transcription. When

X is present it binds to the regulatory protein, converting it to an inactive form that no longer binds to the operator, and thereby turns off transcription.

Mutation 1 alters the regulatory protein such that it no longer can bind to the operator even in the absence of X.

Mutation 5 alters the regulatory protein such that it binds to the operator even when X is present.

As stated in Parts (d) and (e), mutation 9 could be either an operator defect that prevents binding of the regulatory protein, or a promoter defect that prevents binding of RNA polymerase.

1-14 The apparent selection against positive control might be rationalized on the basis of mutational consequences as follows. The most common mutation affecting the control elements of an operon is one that results in a nonfunctional regulatory gene product. In a negatively controlled operon such a mutation causes constitutive synthesis, whereas in a positively controlled operon it causes inability to transcribe the operon under any conditions. In a biosynthetic pathway this sort of mutation would be lethal in the absence of the end product. Thus it would seem more advantagous to an organism to control its biosynthetic operons negatively.

1-15 If the extract contains 6×10^{12} molecules/ml, then the total repressor concentration ($[R_T] = [R] + [RI]$) inside the dialysis sac will be 6×10^{15} molecules/liter or $10^{-8}M$. At $[I] = K_D$, the conditions of the experiment,

$$\frac{[R]}{[RI]} = 1 \text{ or } [R] = [RI] = \frac{[R_T]}{2}$$

Hence the concentration of RI in the sac will be $5 \times 10^{-9}M$, and the bound I will contribute 5 cpm/ml. The level of radioactivity inside the sac therefore will be 1005 cpm/ml, compared to 1000 cpm/ml outside, a difference that would be impossible to detect reproducibly.

Walter Gilbert, who eventually succeeded in purifying the *lac* repressor, overcame the problems of this assay by selecting a mutant strain of *E. coli* with a repressor that had a higher than normal affinity for IPTG and was produced at a level several times higher than the normal 10 molecules per cell.

1-16 The three overlapping palindromes are indicated by lower case letters a, b, and c, respectively, below the sequence. The three corresponding axes of symmetry are indicated by vertical arrows above the sequence. (Adapted from T. Maniatis *et al., Nature,* **250,** 394, 1974.)

1-17   (a)   The probability of finding a particular nucleotide pair at any position in DNA is 1/4. Therefore the probability of these same 16 palindromically matched nucleotide pairs occurring in the same relative positions by chance in a randomly chosen 28-nucleotide-pair sequence is

$$P = (1/4)^{16} = \frac{1}{4.3 \times 10^9}$$

Since there are 4×10^6 possible 28-nucleotide-pair sequences in the DNA molecule (imagine the axis of symmetry being moved along the molecule, one nucleotide pair at a time), the frequency with which this palindrome will occur by chance is, on the average,

$$\frac{4 \times 10^6}{4.3 \times 10^9} = 0.0009 \text{ per } 4 \times 10^6 \text{ nucleotide pairs}$$

(b)   To produce a palindrome of this general character, eight of the 14 nucleotide pairs on one side of the symmetry axis must match appropriately with the nucleotide pairs in the equivalent positions on the other side. If the probability of a match and a mismatch at any position is known, then the probability of eight matches in 14 positions can be calculated using the formula for a binomial distribution. Given a nucleotide pair at any position on one side, the probability of a match at the equivalent position on the other side is 1/4, and the probability of a mismatch is 3/4. The probability of eight matches and six mismatches in 14 positions is given by the appropriate term of the expansion of the binomial $(a + b)^{14}$, in which $a = 1/4$ and $b = 3/4$:

$$P = \frac{14!}{8!\ 6!} (1/4)^8 (3/4)^6$$

$$= 3003 \left(\frac{1}{6.55 \times 10^4}\right) \left(\frac{729}{4096}\right) = 0.00816$$

Thus an average of about one in every 120 randomly chosen 28-nucleotide-pair sequences will contain eight palindromic matches. The average number of such sequences in random sequences of 4×10^6 nucleotide pairs is

$$(0.00816)(4 \times 10^6) = 3 \times 10^4$$

(c)   In order not to exceed the frequency limit of 0.1, all 14 nucleotide pairs on one side must match those on the other. The average frequency of these palindromes will be

$$(1/4)^{14}(4 \times 10^6) = \frac{4 \times 10^6}{2.7 \times 10^8}$$

$$= 0.01 \text{ per } 4 \times 10^6 \text{ nucleotide pairs}$$

The average frequency of 28-nucleotide-pair sequences with 13 matches and one mismatch will be

$$\frac{14!}{13!\ 1!}\ (1/4)^{13}(3/4)(4 \times 10^6)$$

$$= \frac{14 \times 3 \times 4 \times 10^6}{6.7 \times 10^7 \times 4}$$

$$= 0.6 \text{ per } 4 \times 10^6 \text{ nucleotide pairs}$$

Note that all the frequencies in this problem are valid only for random 4×10^6-nucleotide-pair sequences. The sequence of nucleotide pairs in the *E. coli* genome obviously is not random since it is the product of evolutionary selection. If palindromic sequences are, in fact, functionally significant there may be strong selective pressures against their chance occurrence, so that actual frequencies of partial palindromes in *E. coli* are much lower than those calculated.

2 Structure And Organization Of Eucaryotic Chromosomes

The DNA of eucaryotic cells is packaged in organelles called chromosomes, which serve as the vehicles for storage, transmission, expression, and evolution of genetic information. In addition to DNA, chromosomes contain specific proteins that determine their structure and control their functions. This chapter considers the gross structure of chromosomes, the organization of DNA sequences within them, and current ideas on the functional significance of chromosome organization.

ESSENTIAL CONCEPTS

2-1 Chromosomes exist in different structural states at different stages in the cell cycle

(a) The series of stages in normal cell division is called the *mitotic cycle*. During *interphase,* the longest stage of the cycle, the chromosomes are extended so that replication and transcription can occur. Prior to cell division, during the stage called *metaphase,* the chromosomes are contracted about 10,000-fold in preparation for the distribution of genetic information to daughter cells (Essential Concept 3-5). This condensation-extension cycle is accompanied by many changes in structure and function.

Metaphase chromosomes have been studied extensively at the morphological level because they are easily visible in the light microscope after appropriate staining. Interphase chromosomes, which have little visible structure but are functionally more active, have been used for most biochemical studies of chromosomes.

(b) During metaphase the condensed, replicated chromosomes assemble on one plane through the center of the cell, called the metaphase plate. The mitotic *spindle apparatus* then directs the equal distribution of homologous chromosomes to the two daughter cells (Essential Concept 3-4). The cell can be arrested in metaphase with colchicine, an alkaloid drug that blocks assembly of the mitotic spindle apparatus. In the presence of colchicine, an entire cell population can become trapped in metaphase as individual cells proceed through the cell cycle.

(c) The mitotic cycle describes the division of *somatic cells,* which in higher eucaryotes include most of the cells of the organism. The sex cells, or *germ cells,* exhibit an additional mode of division called *meiosis,* which reduces the chromosome content to the haploid number for production of sperm and egg cells. Meiosis and its consequences for inheritance are considered in Chapter 3.

2-2   The karyotype of an organism is the number and morphology of its metaphase chromosomes

(a)   All higher eucaryotes are diploid; that is, each somatic cell has two copies of every chromosome, one of paternal and the other of maternal origin. Homologous chromosomes (homologues) from the metaphase plate that are paired and arranged in order of descending size form a display termed a *karyotype* (Figure 2-1). Each metaphase chromosome consists of two daughter *chromatids* (duplex DNA molecules, see Essential Concept 2-4) joined at a *centromere* (Figure 2-2). Each chromatid is a complete chromosome resulting from replication during the preceding interphase. The centromere is the attachment site for mitotic spindle

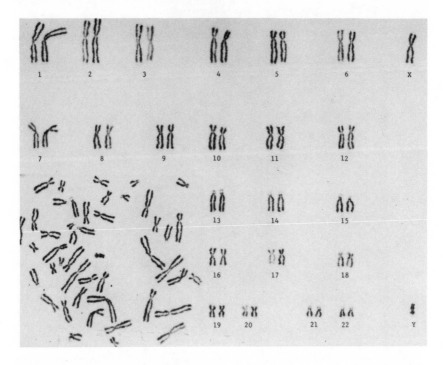

Figure 2-1.   Karyotype of a normal man. The lower left portion of the figure shows chromosomes in metaphase as seen through the microscope. The karyotype has been prepared from another photograph by cutting out each chromosome and pairing it with its homologue. Twenty-two autosomal pairs and the two sex chromosomes X and Y are seen. (From J. German, *Amer. Scientist* **58,** 182, 1970.)

fibers. Its position determines the length of the chromosome arms, a characteristic morphological feature that is helpful in distinguishing metaphase chromosomes from one another. For example, human chromosome 1 has the centromere centrally placed to produce two pairs of long arms, whereas chromosome 10 has one pair of long and one pair of short arms.

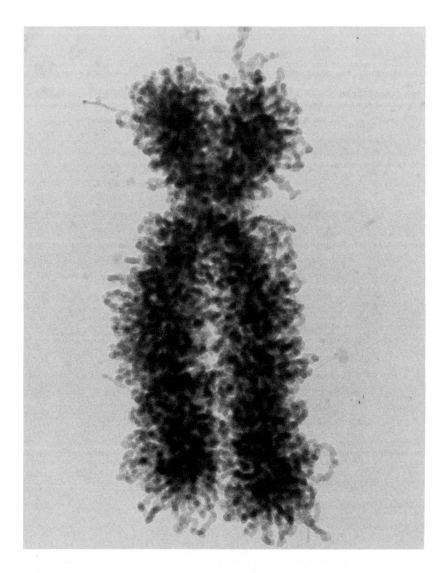

Figure 2-2. Electron micrograph of human chromosome 12 in the metaphase configuration. Two chromatids are clearly visible. The diameter of individual fibers is about 250 Å. (From E. DuPraw, *DNA and Chromosomes,* Holt, Rinehart, and Winston, New York, 1970.)

(b) A variety of staining techniques for metaphase chromosomes, such as quinacrine fluorescence and Giemsa staining, give transverse banding patterns that permit individual chromosomes with otherwise similar morphologies to be distinguished from one another (Figure 2-3). All human chromosomes can be distinguished with these techniques, whereas the ordinary karyotype analysis yields several groups of indistinguishable chromosomes (Figure 2-1).

(c) The two sexes of higher eucaryotic species have different karyotypes. In most species this difference is confined to a specialized pair of chromosomes called *sex chromosomes*. Unlike the remaining chromosomes, which collectively are called *autosomes,* the two sex chromosomes are not homologous.

(1) In mammals as well as in the fruit fly *Drosophila,* a favorite experimental organism for eucaryotic genetics, the two sex chromosomes are designated X and Y. Somatic cells of normal males contain one X and one Y chromosome (XY), whereas somatic cells of normal females contain two X chromosomes (XX). The chromosomes of a female *Drosophila* cell are shown in Figure 2-4.

(2) In a normal female mammal the two X chromosomes in each cell become differentiated irreversibly at an early stage in embryonic development. One remains genetically active, whereas the other becomes condensed and genetically inert. The inactive X chromosome can be visualized in late interphase as a darkly staining

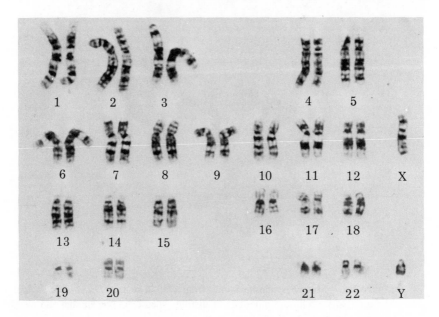

Figure 2-3. Human metaphase chromosomes stained with Giemsa after mild trypsin treatment. (Courtesy Dr. T. C. Hsu.)

particle called a Barr body (Figure 2-5). In male cells the single X chromosome remains active, and no Barr bodies are visible. In cells of abnormal females with three X chromosomes, only one remains active, and two Barr bodies are visible.

(3) In any female embryonic cell, either X chromosome can become inactivated with equal probability. Once this differentiation has occurred, the two X chromosomes retain their respective functional states through subsequent mitotic divisions. It is not known how an X chromosome becomes inactive, or how this inactivation is maintained during division.

(d) The autosomal karyotype among members of a species is invariant, except for infrequent chromosomal abnormalities. In contrast, different species differ in the size, shape, and number of their metaphase chromosomes. For example, whereas humans have 23 pairs of chromosomes per diploid nucleus, *Drosophila* have only four pairs (compare Figures 2-1 and 2-4). Thus the karyotype provides a characteristic "fingerprint" for each individual species.

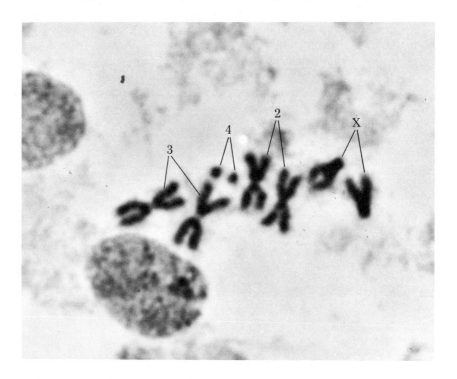

Figure 2-4. Metaphase chromosomes from a larval stage of *Drosophila melanogaster*. The four pairs of chromosomes can be distinguished from one another and identified as shown. (Courtesy Dr. E. B. Lewis.)

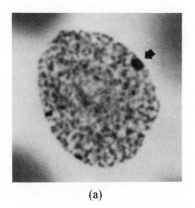

(a)

(b)

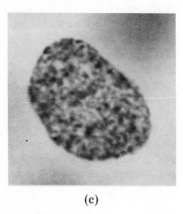

(c)

Figure 2-5.   Barr bodies in interphase nuclei from human cells. (a) Normal female (XX). Arrow indicates Barr body. (b) Abnormal XXX female. Two Barr bodies are visible. (c) Normal male. (From T. Puck, *The Mammalian Cell as a Microorganism*, Holden-Day, San Francisco, 1972.)

(e)   Chromosomal abnormalities can be of three general types: extra or missing individual chromosomes (Figure 2-6), extra or missing portions of a chromosome, or chromosomal rearrangements. The third category includes *translocations* (transfer of a piece from one chromosome onto another chromosome), and *inversions* (reversal in polarity of a chromosomal segment). Individuals whose karyotype deviates from the normal (euploid) state are said to be *aneuploid*. Detectable chromosomal abnormalities occur with a frequency of one in every 250 human births. Abnormalities that involve deletions or additions of chromosomal material alter the gene balance of an organism and generally lead to fetal death or to serious mental and physical defects.

2-3   Most biochemical and biophysical studies of chromosome structure and function have been carried out on interphase chromatin

Chromatin is the name given to the nuclear DNA and its associated materials during interphase, when no obvious chromosomal structure can be seen with the light microscope. Chromatin has been purified by a variety of very gentle techniques, such as isolation and lysis of nuclei followed by differential centrifugation in low-ionic-strength buffers. Unfortunately, the components associated with chromatin vary somewhat, depending upon the method of

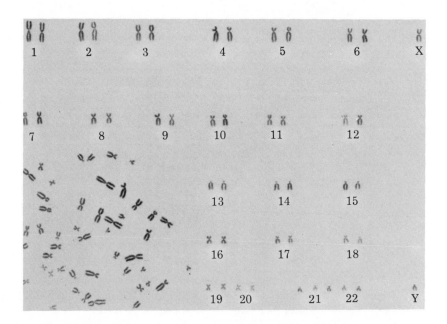

Figure 2-6.   Karyotype of a boy with mongolism showing an extra chromosome 21. This chromosomal abnormality causes mental retardation and numerous gross physical defects. (From J. German, *Amer. Scientist* **58,** 182, 1970.)

isolation. This observation raises the question of whether the composition of isolated chromatin is the same as that of *in vivo* chromatin. Despite this complication, most investigators agree that mammalian chromatin consists almost entirely of DNA and two classes of proteins, histones and nonhistone chromosomal proteins, in a weight ratio of approximately 1:1:1.

2-4   A single duplex DNA molecule forms the structural backbone of each chromosome

(a)   Duplex DNA molecules isolated from *Drosophila* and yeast chromatin have lengths corresponding to the DNA content of single chromosomes. This result indicates that each chromosome contains only a single very long DNA molecule. The DNA content of the largest human chromosome (#1) is approximately 2.7×10^8 nucleotide pairs, whereas the content of the smallest (#22) is approximately 0.5×10^8 nucleotide pairs.

(b)   In ascending the evolutionary scale of animals, the DNA content per haploid set of chromosomes generally increases (Figure 2-7). For example, *E. coli* has 4.5×10^6 nucleotide pairs in its genome, *Drosophila* has 1.4×10^8 nucleotide pairs, and man has 3.2×10^9 nucleotide pairs. If 1000 nucleotide pairs are sufficient to code for an average polypeptide, then *E. coli* has sufficient informa-

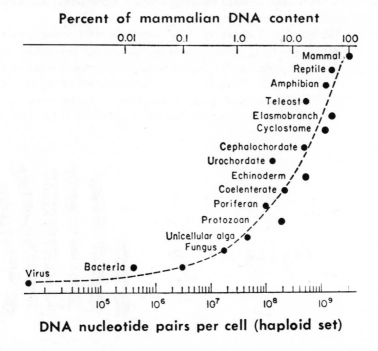

Figure 2-7.   Minimum haploid DNA content in species at various levels of organization. (From R. Britten and E. Davidson, *Science* **165**, 349, 1969.)

tion for 4500 proteins, *Drosophila* for 1.4×10^5 proteins, and man for 3.2×10^6 proteins. To many biologists the numbers for higher organisms appear unexpectedly large. They raise a fundamental question: How much of the DNA in higher organisms actually is used to direct RNA and protein synthesis?

(c)   In certain families of plants and animals, such as the Ranunculacea (buttercups) and some amphibia, the DNA content per haploid nucleus (C-value) can differ by two orders of magnitude between closely related species with similar karyotype and morphology (Figure 2-8). This observation, termed the C-value paradox, also raises questions as to the function of the extra DNA. In contrast, other classes of animals, such as mammals, all have very nearly the same haploid DNA content, although their karyotypes differ greatly.

2-5   DNA in chromosomes is complexed with histones, nonhistone proteins, and a small amount of RNA

(a)   The histones are sufficiently similar in all organisms that they can be divided into five major classes according to their amino acid compositions, chromatographic and electrophoretic properties, and amino acid sequences. Some heterogeneity exists within each class due to minor sequence variations and side-chain modifications, but the classes in any given tissue or organism usually can be distinguished from one another. Some properties of the histones from calf thymus are given in Table 2-1.

The amino acid sequences of the histones reveal two important patterns.

(1)   Most basic amino acids in any histone are crowded into one half of the molecule, leaving the amino acid composition of the other half relatively "normal." This distribution has functional implications since the basic portion presumably interacts with the negative charges on DNA, whereas the other portion is free to interact with

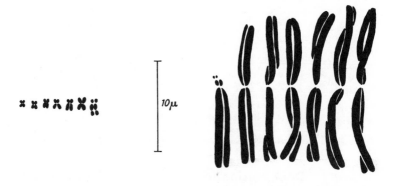

Figure 2-8.   Karyotypes of two species of buttercups *(Ranunculaceae),* which differ 80-fold in haploid DNA content. (Adapted from K. Rothfels *et al., Chromosoma* **20,** 54, 1966.)

Table 2-1. Nomenclature and characteristics of histones from calf thymus.

Description	Alternative nomenclatures			Molecular weight
Lys-rich	I	f1	H1	21,000
Slightly Lys-rich	IIb1	f2b	H2b	13,700
	IIb2	f2a2	H2a	14,500
Arg-rich	III	f3	H3	15,300
	IV	f2al	H4	11,300

other histones, with nonhistone proteins, and with the neutral and hydrophobic regions of the DNA helix.

(2)   Certain histones are highly conserved in their amino acid sequences. For example, histone IV differs in only two of 102 residues and histone III differs in four of 135 residues between calf and pea, two species that diverged almost a billion years ago. Histones IIb1 and IIb2 are less highly conserved (eight differences out of 125 residues in IIb1 and three differences in the first 22 residues of IIb2 between trout and calf), and histone I appears quite variable (eight differences in the first 34 residues between rabbit and calf). Presumably the highly conserved histones are maintained invariant by natural selection because of the precise requirements of their interactions with DNA and with each other.

(b)   The nonhistone chromosomal proteins are heterogeneous. These proteins include chromosomal enzymes such as DNA and RNA polymerases, nucleases, chromosomal structural proteins, and regulators of gene expression. Little is known about these proteins. Typically they are more acidic than histones, and probably include more than 100 different polypeptides ranging in molecular weight from 10,000 to 300,000 (Figure 2-9). Since these proteins are designated as chromosomal proteins on the basis of their presence in isolated chromatin, the question must be raised as to whether all of them are associated with chromosomes *in vivo* or whether some are present only as artifacts of the chromatin isolation procedure.

(c)   RNA is a minor component of chromosomes. RNA in the process of transcription (nascent RNA) is associated with isolated chromatin. Other low-molecular-weight RNA found in chromatin may play some structural or regulatory role, although evidence for these suppositions is lacking. No macromolecules other than protein and nucleic acids are found in chromatin in significant quantity.

2-6   A complex of eight histone molecules and about 200 nucleotide pairs of DNA is a basic repeating unit of chromosome structure

(a)   DNA from most, if not all, eucaryotes is associated with a nearly equal

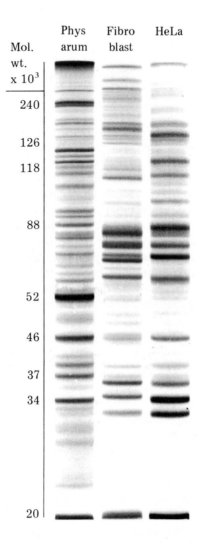

Figure 2-9. Major components of the nuclear and nonhistone chromosomal proteins from three different eucaryotes compared by an electrophoretic technique (acrylamide gel in the presence of sodium dodecyl sulfate) that separates proteins according to molecular weight. The nonhistone chromosomal proteins constitute a collection of many different sized polypeptides varying over a wide range of concentrations, whereas the histones fall into four or five major size classes when examined by a similar technique. (From W. Lestourgeon *et al.*, *Biochem. Biophys, Acta,* in press, 1975.)

weight of histone, suggesting that the molar relationships of DNA and histone are highly conserved throughout evolution. The molar composition of calf chromatin corresponds to about one molecule of histone I and two molecules each of histones IIb1, IIb2, III, and IV per 200 base pairs of DNA. This composition corresponds to a weight ratio of 1:1 for DNA to histone.

(b) Cross-linking and ultracentrifugation experiments have shown that histones III and IV form a tetramer, III_2IV_2. Histones IIb1 and IIb2 form oligomers of composition $(IIb1\text{-}IIb2)_n$ but unknown structure, both in solution and in association with chromatin. A mixture of these two kinds of oligomers and free DNA forms a complex that gives x-ray diffraction patterns identical to those obtained from intact chromatin. Histone I is not required to obtain these patterns. In addition, the amount of histone I complexed with DNA can vary from one organism to another. Accordingly, its role in chromosome structure is uncertain.

(c) Cleavage of chromatin with certain nucleases suggests a repeating structure of about 200 nucleotide pairs. For example, a staphylococcal nuclease can cleave about 80% of the DNA of rat chromatin into pieces of this size.

(d) The x-ray patterns of native and reconstituted chromatin indicate a regular structure with a repeat interval of 100 Å along the chromatin fiber. Generally, electron micrographs show fiber diameters of about 100 Å. Close-packed chromatin is about 45% by weight histone and DNA. Assuming a density of 1.5 g/cm³ for chromatin, it can be calculated that a unit 100 Å in length and 100 Å in diameter contains 2.8×10^5 daltons of chromatin or about 1.4×10^5 daltons each of DNA and histone. These quantities are equivalent to 230 nucleotide pairs of DNA and 2.3 molecules each of histones IIb1, IIb2, III, and IV. Thus both chemical and physical analyses support the conclusion that the fundamental repeating structural unit of chromatin is likely to consist of a histone III_2IV_2 tetramer, two molecules of each IIb histone associated in an unknown manner, and about 200 DNA nucleotide pairs. Within this unit the DNA must be locally condensed, since the extended length of 200 nucleotide pairs would be 680 Å.

(e) Chromatin fibers must be able to coil or fold extensively, as in metaphase chromosomes, and accordingly must be flexible. Electron micrographs of chromatin fibers show a fine structure resembling beads on a string, suggesting a flexibly jointed chain of repeating units (Figure 2-10). One plausible model suggests that the $III_2 IV_2$ tetramer plus a IIb1-IIb2 tetramer could form the repeating unit with histone I determining the spacing of octamers along the length of the chromatin fiber. Thus the fiber might consist of tightly condensed DNA and associated protein alternating with more extended DNA and associated protein, to provide flexibility.

2-7 Chromosomes show regional differentiation into heterochromatic and euchromatic regions

(a) Interphase chromosomes are visible in the light microscope when appropriately stained. Under these conditions, two distinct kinds of organization, termed *heterochromatic* and *euchromatic,* can be seen in different chromosomal regions. Heterochromatic regions are condensed and stain darkly. These same regions replicate late in the cell cycle and are thought to represent genetically inert chromatin. Euchromatic regions are more dispersed and stain lightly. These regions replicate early in the cell cycle and have the potential for genetic expression.

(b)   There are two types of heterochromatin.

   (1)   *Constitutive* heterochromatin retains its heterochromatic state in virtually all cell types and stages. The centromeric regions of all chromosomes consist of constitutive heterochromatin.

   (2)   *Facultative* heterochromatin may be either heterochromatic or euchromatic in the same organism depending upon cell type or stage of development. The inactive X chromosomes in female mammalian somatic cells are examples of facultative heterochromatin.

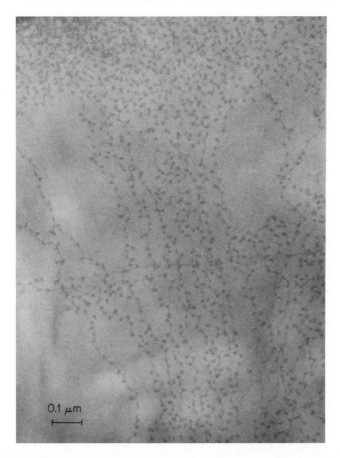

0.1 μm

Figure 2-10.   Electron micrograph of chromatin fibers from a chicken erythrocyte nucleus showing a structure resembling beads on a string. (Courtesy Dr. D. E. Olins.)

2-8 **Polytene chromosomes are aligned bundles of identical interphase DNA molecules that allow direct visualization of functional genetic units**

(a)   Polytene chromosomes are found in certain tissues of several organisms. The polytene chromosomes from *Drosophila* salivary glands have been studied extensively. These chromosomes arise from the chromosomes of diploid nuclei by successive duplications that differ from those of the normal mitotic cycle in the following respects. Homologous chromosomes pair and remain paired throughout a polytenization. The chromosomes no longer undergo the normal mitotic cycle of coiling and uncoiling. Following replication, daughter chromosomes do not segregate but remain paired with each other. The nuclear membrane remains intact following replication, and there is no cell division. Nine to ten cycles of such replication result in a nucleus containing the haploid number of giant polytene chromosomes, each composed of 1000-2000 identical extended DNA molecules in a multistranded cable.

(b)   Along the linear axis of each polytene chromosome, bands or *chromomeres* are visible in the light microscope (Figure 2-11). Presumably these bands repre-

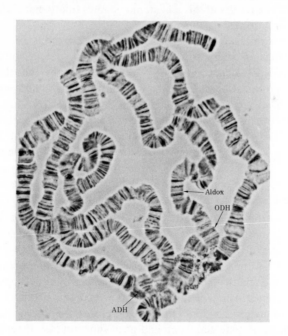

Figure 2-11.   Polytene chromosomes from the salivary gland of *Drosophila melanogaster*. The four chromosomes are closely connected at their centromeric regions to form a compact chromocenter, from which the two arms of each chromosome protrude. The structural genes for the three enzymes aldehyde oxidase (aldox), alcohol dehydrogenase (ADH), and octanol dehydrogenase (ODH) have been associated with specific bands by combined biochemical and cytogenetic techniques. (Courtesy Dr. H. Mitchell.)

sent units of organization on individual chromatids that become visible when they are aligned in large numbers. Each polytene chromosome has a characteristic pattern of bands and interbands that is constant for all the polytene tissues of the organism and specific for the species. The polytene chromosomes of *Drosophila melanogaster* exhibit about 5000 bands. The bands contain 95% of the chromosomal DNA; the interband regions contain 5%. The DNA content of single bands ranges from about 3×10^3 to 3×10^5 nucleotide pairs.

(c)   Nucleic acid reassociation studies have demonstrated that only the euchromatic regions of *Drosophila* chromosomes undergo polytenization. In addition, certain gene families with multiple members, such as the ribosomal genes, appear to undergo only partial polytenization. Therefore the duplication process for polytenization must be complex.

(d)   By a variety of genetic techniques, mutant phenotypes have been correlated with changes in individual polytene bands (Figure 2-11). In the most complete such study, a region of the X chromosome of *Drosophila* containing twelve bands was saturated with 160 lethal and semilethal mutations, and these mutations were shown to fall into only 12 units of function or complementation groups. The results of this and similar studies imply that each band represents a single unit of genetic function.

This implication presents a paradox. The units of genetic function in procaryotes are single structural genes, that is, stretches of DNA that code for single polypeptides. If the same is true in eucaryotes, then the one function-one band hypothesis suggests that there are only 5000 structural genes in *Drosophila,* less than twice the number thought to be present in *E. coli.* Intuitively, this number seems too low. Moreover, the average band contains about 30,000 nucleotide pairs of DNA, sufficient information to code for 30 average proteins. If there is only one structural gene per band, then the function of the remaining DNA is unclear. Both considerations argue for more than one structural gene per chromomere, in apparent conflict with the genetic results. Some attempts to resolve this conflict are presented in Essential Concept 2-15.

(e)   Swollen bands called puffs appear at specific loci on polytene chromosomes at particular stages in embryonic development. Each puff represents a site of RNA synthesis at which, presumably, DNA fibers uncoil so that they can be transcribed. Ecdysone, an insect moulting hormone, induces puff formation in particular bands within 30 minutes of direct application to polytene chromosomes in certain dipteran larvae, whereas actinomycin D, an inhibitor of RNA polymerase, inhibits puffing. Therefore chromosome puffs allow direct visualization of gene activation and gene function (see Essential Concept 5-9).

2-9   Lampbrush chromosomes are paired meiotic chromosomes that also permit direct visualization of functioning chromomeres

(a)   The lampbrush chromosome consists of two homologous chromosomes in the extended state, held together by one or more *chiasmata* (chromosomal crossovers, Figure 2-12). Each homologue contains two sister duplex DNA mole-

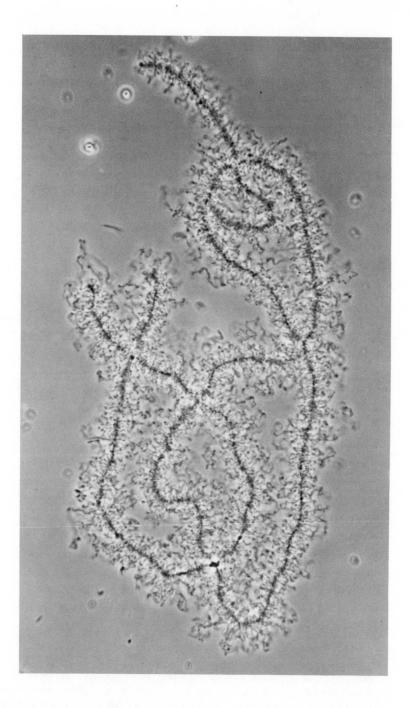

Figure 2-12. A lampbrush chromosome from an oocyte of the newt *Triturus,* showing the two homologues held together by several chiasmata. The fuzzy appearance is due to the characteristic lateral loops. (Courtesy Dr. J. Gall.)

(a)

(b)

(c)

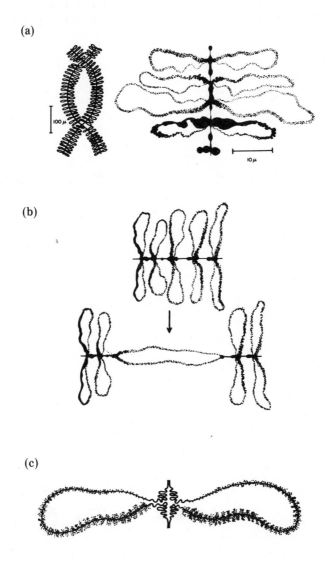

Figure 2-13. Diagrammatic view of lampbrush chromosomes. (a) The two homologous meiotic chromosomes (left) are joined together by two chiasmata. A portion of the central chromosome axis (right) shows that two loops with identical morphology emerge at a given point, evidence that each chromosome already has split into two chromatids. (b) Accidental stretching of a chromosome reveals the continuity of the loop axis with the central axis (compare with Figure 2-2). (c) A single loop pair showing the single DNA molecules upon which RNA chains are being made. (From J. Gall, *Brookhaven Symp. Biol.* **8,** 17, 1955.)

cules. A symmetrical pair of very thin loops emanates from each of the hundreds of chromomeres that are distributed along each chromosome (Figure 2-13). Lampbrush chromosomes are found in many diverse species, including man.

(b)   The lampbrush loops, which are sites of active RNA synthesis, presumably represent functional units of DNA that are extended to allow transcription. There are about 5000 loops per haploid set of lampbrush chromosomes in certain amphibia, the same as the number of bands seen in polytene chromosomes of *Drosophila*. Some geneticists suggest that loops and bands represent similar functional genetic units on meiotic and mitotic chromosomes, respectively.

2-10   DNA sequences can be grouped into three general classes based on frequency of occurrence in the genome

(a)   If the DNA of an organism is sheared into small fragments of several hundred nucleotide pairs, denatured, and then incubated under conditions optimal for reassociation of complementary strands, the rate of reassociation for a particular sequence will be related to the number of copies of that sequence in the genome. Sequences represented only once in the genome (unique sequences) will hybridize slowly compared to sequences that are present in many copies. Analysis of DNA-DNA reassociation kinetics is explained in the Appendix to this chapter.

(b)   DNA-DNA reassociation experiments show that all eucaryotes have repeated DNA sequences. In contrast, procaryotes have almost exclusively unique DNA sequences. Eucaryotic sequences can be divided somewhat arbitrarily into three general frequency classes, termed highly repetitive (also called satellite DNA), middle-repetitive, and unique. Different eucaryotic organisms contain differing amounts of these frequency classes. The copy frequencies and range of percentages of these three classes are given in Table 2-2. Often the

Table 2-2. Three frequency classes of DNA sequences found in eucaryotes.

Frequency class of DNA	Percentage of genome	Number of copies per genome	Examples
Unique	10%-80%	1	Structural genes for hemoglobin, oval-bumin, and silk fibroin
Middle-repetitive	10%-40%	10^1-10^5	Genes for rRNA, tRNA, and histones
Highly repetitive	0%-50%	$>10^5$	Satellite DNA sequences of 5-300 nucleotides

middle-repetitive classes sometimes can be divided further into slow, inter-mediate, and fast components. Table 2-3 shows the abundance of the various classes in each of several animals.

(c)   DNA fragments that contain inverted repeats or palindromic sequences [Essential Concept 1-8(c)] are exceptions to the generalization in (a). These fragments reassociate very rapidly to form hairpin structures with intramolecular complementarity. Since this reaction is monomolecular, its rate is independent of concentration. Consequently, the rapid reassociation of these sequences does not indicate that they are highly repetitive. Clusters of palindromes appear to be scat-tered throughout eucaryotic DNA. Their function, if any, is unknown.

Table 2-3. Frequency components of several animal DNA's. (Adapted from E. Davidson and R. Britten, *Quart. Rev. Biol.* **48**, 565, 1973.)

Species and component	Copies per genome	Percentage of genome	Complexity[a]
Nassaria obsoleta (snail)			
Unique	1	38%	1.1×10^9
Middle-repetitive			
Slow	20	12%	1.7×10^7
Intermediate	1,000	15%	4.5×10^5
Highly repetitive and/or palindromes	—	18%	—
Calf			
Unique	1	55%	1.5×10^9
Middle-repetitive	60,000	38%	1.7×10^4
Highly repetitive	1,000,000	2%	60
Highly repetitive and/or palindromes	—	3%	—
Xenopus laevis			
Unique	1	54%	1.6×10^9
Middle-repetitive			
Slow	20	6%	1.5×10^7
Intermediate	1,600	31%	6.0×10^5
Fast	32,000	6%	6.0×10^3
Highly repetitive and/or palindromes	—	3%	—
Strongylocentrotus purpuratus (sea urchin)			
Unique	1	38%	3.0×10^8
Middle-repetitive			
Slow	20-50	25%	1.0×10^7
Intermediate	250	27%	1.0×10^6
Fast	6,000	7%	1.3×10^4
Highly repetitive and/or palindromes	—	3%	—

[a]See Appendix, Section A2-3.

2-11 **Highly repetitive DNA sequences are located in regions of centromeric heterochromatin**

(a) Highly repetitive or satellite DNA was isolated first from the mouse (*Mus musculus*) as a minor (10%) component that differed in buoyant density from the bulk of the DNA. DNA-DNA reassociation experiments suggest that mouse satellite DNA is composed of one million very similar copies of a nucleotide sequence about 300 base pairs in length. These DNA sequences are not transcribed into RNA *in vivo*.

(b) Satellite DNA sequences are clustered in the heterochromatic centromeric region of all mouse chromosomes, except possibly the Y, as shown by *in situ* hybridization. In this technique satellite DNA sequences are transcribed *in vitro* with *E. coli* RNA polymerase, using highly radioactive nucleotides. The labeled RNA molecules then are allowed to associate (hybridize) with chromosomes *in situ,* in cells that have been arrested at metaphase, lysed, and exposed to mild denaturing conditions. The resulting preparation is examined by combined light microscopy and autoradiography to determine the locations and extents of satellite DNA sequences on the chromosomes (Figure 2-14).

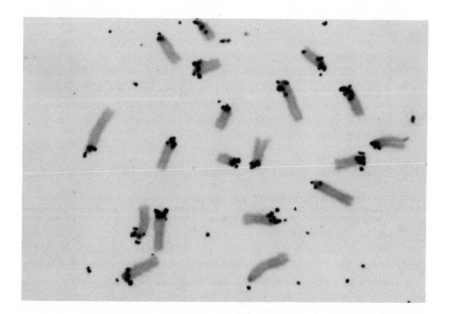

Figure 2-14. Chromosomes of the mouse hybridized *in situ* with RNA copied from mouse satellite DNA. Only the centromeric heterochromatin has labeled. (Courtesy Dr. J. Gall and Dr. M. L. Pardue.)

(c)   The function of satellite DNA is unknown. It has been found in many eucaryotes and shown to exhibit the following characteristics.

(1)   It is composed of one or more families of precisely repeated sequences.

(2)   The sequences of a given family are clustered at one chromosome locus.

(3)   Satellite DNA usually is located in regions of constitutive heterochromatin.

(4)   Generally, it is concentrated close to centromeres or chromosome ends (telomeres).

(5)   It is transcribed to a very small extent, if at all, within the cell.

(6)   It evolves very rapidly. For example, other species of mice have sequences only distantly related to the satellite DNA found in *Mus musculus*.

2-12   Middle-repetitive sequences are interspersed among unique sequences

(a)   Middle-repetitive DNA is a broad and heterogeneous class that contains some families of similar sequences as well as other families with divergent sequences. In contrast to satellite DNA, some middle-repetitive sequences are transcribed.

(b)   Analysis of sequence organization in the DNA of two organisms, the African toad *Xenopus* and the sea urchin *Strongylocentrotus,* has shown similar patterns. Much of the genome consists of interspersed repetitive and unique sequences. The repetitive and unique sequences in these DNA's are about 300 nucleotide pairs and one thousand nucleotide pairs in length, respectively. This striking organizational pattern is consistent with the hypothesis that interspersed repetitive sequences somehow control the transcription of the contiguous unique sequences (see Essential Concept 5-14).

2-13   Several known proteins are coded by unique DNA sequences

(a)   Messenger RNA for a few specific proteins (hemoglobin from mouse and duck, silk fibroin from silk moth, and ovalbumin from chicken) has been labeled radioactively and isolated. In some cases a radiolabeled DNA transcript (cDNA) has been made using the purified mRNA as a template for RNA-directed DNA polymerase. From the kinetics with which the labeled mRNA or cDNA associates with complementary DNA sequences, the frequency classes of these sequences can be determined (see Appendix, Chapter 5). The results indicate that each of these proteins is coded by a unique DNA sequence.

(b)   In related experiments the hybridization kinetics of total mRNA and denatured DNA from several kinds of cells have been analyzed. The results show that most of the mRNA in cells of sea urchin embryos, a mouse tissue culture line (L cells), a human tissue culture line (HeLa cells), and the cellular slime mold

Dictyostelium hybridize with unique DNA sequences. Accordingly, most structural genes in eucaryotes are probably present in only one copy per haploid genome.

(c)   The ease of isolation of rRNA, tRNA, and the corresponding DNA sequences greatly simplified the analysis of rRNA and tRNA genes and their mapping by *in situ* hybridization. Recently, an important new technique has been developed that allows the isolation in quantity of other sequences, randomly selected from the *Drosophila* genome. A preparation of *Drosophila* DNA, sheared to a desired size range, is mixed with DNA of an *E. coli* episome (plasmid) carrying a gene that confers resistance to the drug tetracycline. The mixture is subjected to a series of enzymatic techniques that promotes end-to-end splicing of DNA fragments to one another. Some of the molecules so produced are circular hybrids that include the bacterial plasmid genes as well as a segment of *Drosophila* DNA. *E. coli* then are exposed to the treated DNA preparation under conditions where the cells take up DNA, and tetracycline-resistant bacterial clones are isolated and grown into large cultures. The cells in each such culture carry a unique plasmid, which may include a segment of *Drosophila* DNA. The plasmid DNA can be isolated easily and employed for *in situ* hybridization to polytene chromosomes and for other experiments. This technique can be used to propagate individual sequences from any population of DNA molecules and undoubtedly will find broad application.

2-14   The genes for rRNA, tRNA, and histones occur as clusters of repeated sequences

(a)   The genes for tRNA and rRNA in the African toad *Xenopus* are found in repeating sequences of the general form [spacer:RNA gene]$_n$. One such repeating unit contains the genes for 18S and 28S rRNA. Another contains the gene for 5S rRNA, and others contain individual tRNA genes. These three kinds of repeating units occur in separate clusters in the genome. They represent the best-known structures of specific genes.

   (1)   The DNA that contains the 18S and 28S rRNA genes of *Xenopus* can be isolated easily because it differs from the rest of the genome in GC content and therefore in buoyant density. Some characteristics of these genes are given in Table 2-4. From *in situ* hybridization they

Table 2-4. Characteristics of *Xenopus laevis* rRNA genes. (Adapted from D. Brown and K. Sugimoto, *Cold Spring Harbor Symposia on Quantitative Biology* **38**, 501, 1973.)

Percent of total DNA:	0.2
Average number of repeats:	450
Chromosomal location:	Nucleolar organizer (one chromosome)
Repeat length:	13,000 nucleotide pairs
Ratio of gene length to spacer length:	1:1

are known to be clustered at a single chromosomal site, called the nucleolar organizer, on one of the 18 chromosomes of the *Xenopus* haploid set. The tandemly linked 18S and 28S rRNA genes and a spacer of about the same size [Figure 2-15(a)] constitute a unit that is repeated 450 times in the haploid genome. Reassociation experiments show that these units are very similar to one another.

(2)   The 5S rRNA genes are present in clusters on the telomeres of the long arms of at least 15 of the 18 *Xenopus* chromosomes. Some characteristics of the 5S rRNA genes are given in Table 2-5, and the repeat units are depicted in Figure 2-15(b). The 5S rRNA genes show some heterogeneity. *Xenopus mulleri* synthesizes at least two types of 5S rRNA that differ by seven of 120 nucleotides. The spacer sequences also are heterogeneous, based on hybridization experiments

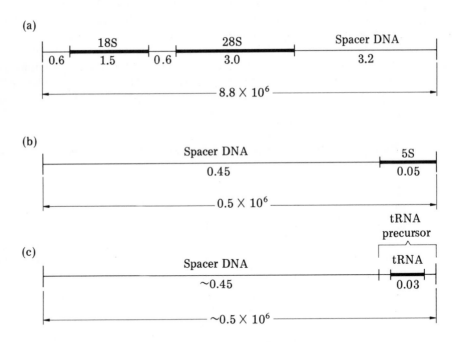

Figure 2-15. Repeating units for rRNA and tRNA. Numbers indicate approximate molecular weights of RNA. (a) Schematic diagram of one repeating unit of rRNA genes from *Xenopus laevis*. The 18S and 28S gene regions are denoted by heavy lines. These genes are co-transcribed as a 45S precursor, which is processed before transport to the cytoplasm (see Essential Concept 5-11). (b) Schematic diagram of one repeating unit of 5S rRNA genes from *Xenopus laevis*. (c) Schematic diagram of a tRNA gene repeat unit of *Xenopus laevis*. [Parts (a) and (b) adapted from D. Brown and K. Sugimoto, *Cold Spring Harbor Symposia on Quantitative Biology* **38**, 501, 1973. Part (c) adapted from S. Clarkson and M. Birnstiel, *Cold Spring Harbor Symposia on Quantitative Biology* **38**, 451, 1973.]

Table 2-5. Characteristics of *Xenopus laevis* 5S DNA. (Adapted from D. Brown and K. Sugimoto, *Cold Spring Harbor Symposia on Quantitative Biology* **38**, 501, 1973.)

Percent of total DNA:	0.7
Average number of repeats:	24,000
Chromosomal location:	Most telomeres
Repeat length:	700 nucleotide pairs
Ratio of gene length to spacer length:	1:6

and preliminary sequence analysis. Whether the individual units of a given 5S rRNA gene cluster are identical is not known.

(3) Preliminary studies on the tRNA genes of *Xenopus* are summarized in Table 2-6. Distinct tRNA gene clusters, each presumably coding for one species of tRNA, occur at many different chromosomal sites. The number of tRNA genes per cluster ranges from 90 to 330 in the clusters studied. A schematic diagram of the repeating unit is given in Figure 2-15(c).

(b) The genes for each of the five histones in *Drosophila* and sea urchin appear to be present in tandemly repeating clusters of 100 and 400 copies, respectively. Each histone appears to be homogeneous by the conventional criteria of protein chemistry (although the results do not rule out a 5%-10% microheterogeneity), and the gene copies that code for each histone are very similar if not identical as judged by hybridization experiments. *In situ* hybridization of histone mRNA with *Drosophila* polytene chromosomes suggests that at least three classes of histones are coded by genes in a single region that includes two or three faint bands. These results imply that the histone genes also occur in tandemly linked clusters.

2-15 The functional implications of chromosome organization are still in dispute

(a) Three general interpretations of chromosome organization have been proposed during the past several years (Figure 2-16). These models are attempts to explain the discrepancy discussed in Essential Concept 2-

Table 2-6. Gene and cluster repeats of tRNA genes in *Xenopus laevis*. (Adapted from S. Clarkson and M. Birnstiel, *Cold Spring Harbor Symposia on Quantitative Biology* **38**, 451, 1973.)

Type of tRNA gene	Number of genes	Number of tRNA clusters
Total tRNA	7800	43
tRNA^{Asp}	180	~1
tRNA^{Leu}	350	~4
tRNA^{Val}	240	~1
tRNA$_1^{Met}$	330	~1
tRNA$_2^{Met}$	170	~1

9(d) between haploid DNA content and apparent number of genetic units of function, and to account for the large variation in DNA content between some closely related species.

(1) The junk model postulates that a small proportion of the genome carries genetic information, while the remaining DNA is important only for chromosome structure. In apparent support of this hypothesis are arguments from population genetics that if the number of essential genes in an organism exceeded a few thousand, the rate of accumulation of deleterious mutations (genetic load) would be high enough to exterminate the species. According to this model, each genetic unit of function consists of a single structural gene and the appropriate amount of junk DNA, which can vary between closely related species without affecting their genetic makeup.

(2) The master-slave model postulates that functional genetic units consist of a master gene (M), and many identical slave genes (S), each of which is corrected against the master gene at each meiotic DNA replication. Thus heritable mutations occur only in the master gene, and the entire cluster evolves together. According to this model, differences in DNA content between closely related species could reflect simple differences in the average number of slave genes per master.

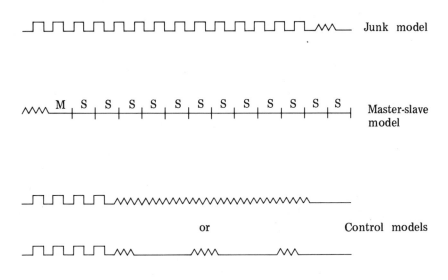

Figure 2-16. Three models for organization of a chromomere and its interchromomeric region. ⎍⎍ indicates junk DNA; ∿∿∿ indicates control DNA; and ———— indicates structural genes. M indicates a master gene and S a slave gene (see text).

(3)   The regulatory model postulates that most DNA in each genetic unit of function represents control elements that somehow regulate the expression of one or a few structural genes. This model offers no convenient explanation for large variations in DNA content between closely related species.

(b)   Several observations place constraints on these models.

(1)   The finding that most mRNA molecules are transcripts of unique DNA sequences argues strongly against the master-slave hypothesis. Even in related amphibia whose haploid DNA contents differ by a factor of ten, the mRNA molecules from comparable tissues hybridize with unique DNA sequences.

(2)   Functional mRNA transcripts of about 14,000 gene-equivalents of unique DNA are found in one particular stage of sea urchin embryo and about 4000 and 35,000 gene-equivalents are found in *Drosophila* and cultured human cells, respectively. These data argue strongly against the junk model and any other model that predicts that eucaryotes require only a few thousand structural genes.

(3)   In mouse brain tissue, about 10% of the unique DNA sequences, corresponding to 300,000 gene-equivalents of DNA are transcribed. In kidney and liver, about 3% of the unique sequences are transcribed. Although these transcripts are known to represent more than functional mRNA, such experiments also argue that the number of structural genes in eucaryotes is more than a few thousand.

(c)   Although most biologists feel that current evidence favors the regulatory model, two paradoxes remain unsolved.

(1)   Genetic evidence for a one-to-one correspondence between gene functions and chromomeres in *Drosophila* has not been reconciled with the demonstration that eucaryotes appear to utilize more than a few thousand structural genes. At least a partial explanation may be that different bands contain different numbers of structural genes and different numbers of control elements. Accordingly, a band for histones may contain 100 genes; the band for a giant polypeptide in the salivary gland may represent a single gene; and other bands may contain intermediate numbers of structural and regulatory genes.

(2)   The large differences in haploid DNA content between some closely related species (C-value paradox) are not yet understood. The expansion in content from one such species to the other does not reflect simple multiplication of the entire genome, since the amounts of repetitive DNA in these organisms differ more than the amounts of unique DNA. There is some evidence that certain multigene families, including the genes for rRNA and tRNA, are expanded in proportion to total genome expansion, but these families can account for only a small fraction of the differences observed. Thus neither the

function of most of the expanded sequences nor the selective value of such expansion is clear.

2-16 Additional concepts and techniques are presented in the Appendix and the Problems section

(a) Analysis of nucleic acid reassociation kinetics. Appendix.

(b) Analysis of frequency classes in eucaryotic genomes. Problems 2-18, 2-19, 2-20, 2-21, 2-22, and 2-26.

(c) Thermal denaturation analysis of reassociated DNA. Problems 2-23 and 2-24.

(d) Effects of *criterion* on analysis of frequency classes. Problem 2-25.

(e) Limitations of *in situ* hybridization. Problem 2-28.

(f) Analysis of gene organization by reassociation experiments. Problem 2-29.

APPENDIX: ANALYSIS OF NUCLEIC ACID REASSOCIATION KINETICS

A2-1 Separated complementary strands of nucleic acids can reassociate to form base-paired duplex structures

(a) Kinetic analysis of nucleic acid reassociation is a valuable technique for studying the structure, organization, and expression of eucaryotic genomes. The technique, described here for DNA-DNA reassociation, also can be used to study DNA-RNA and RNA-RNA association (see Appendix, Chapter 5). DNA-RNA association is referred to as hybridization.

(b) The rate at which complementary sequences reassociate depends upon four parameters: (1) the concentration of cations, which decrease the intermolecular repulsion of negatively charged DNA strands, (2) the incubation temperature, which is optimal for reassociation at about 25°C below the melting temperature (T_m) of the original duplex DNA (see WWBH Essential Concept 16-3), (3) DNA concentration, which determines the frequency of intermolecular collisions, and (4) the size of the DNA fragments.

(c) Reassociation studies generally are carried out as follows.

    (1) Purified DNA is sheared into small fragments by passage through a small orifice at high pressure or by high-speed blending. Uniform fragment sizes of 3×10^2 to 10^5 nucleotide pairs can be produced by controlled shearing.

    (2) The fragments are denatured by heating briefly to 100°C and then cooling rapidly. This procedure separates complementary strands

without breaking a significant number of internucleotide bonds.

(3)   The dissociated fragments are incubated in a standard phosphate buffer (generally $0.18N$ monovalent cation) at a DNA concentration and temperature that will give an appropriate rate of reassociation (*annealing* conditions).

(4)   The extent of reassociation as a function of time is measured by one of the three following techniques, which distinguish between double- and single-stranded DNA: *Nuclease digestion*. Using a nuclease that catalyzes hydrolysis of single-stranded but not double-stranded DNA (such as S_1 nuclease from the mold *Aspergillus*), samples can be digested to completion and the amount of nuclease-resistant material determined. *Hypochromicity*. Since double-stranded DNA absorbs less ultraviolet light than single-stranded DNA at the same concentration, the degree of reassociation is directly proportional to the decrease in absorbance of 260 nm light (hypochromicity) as the reaction proceeds. *Hydroxyapatite chromatography*. Since double-stranded DNA adsorbs to hydroxyapatite under conditions where single-stranded DNA does not, reassociation can be followed by passing samples through a column and determining the fraction of the DNA that adsorbs. Hydroxyapatite chromatography is useful also as a preparative technique for separating rapidly reassociating from slowly reassociating fragments.

A2-2   Reassociation follows bimolecular reaction kinetics

DNA reassociation is a bimolecular second-order reaction. The rate of disappearance of single strands is given by the expression

$$-\frac{dC}{dt} = kC^2 \tag{A2-1}$$

in which C is the concentration of single-stranded DNA in moles of nucleotide/liter, t is time in seconds, and k (liters/mole seconds) is a second-order rate constant. The value of k depends upon cation concentration, temperature, fragment size, and the sequence complexity of the DNA population (Section A2-3). Equation A2-1 can be rearranged to

$$-\frac{dC}{C^2} = kdt$$

and integrated from initial conditions $t = 0$ and $C = C_0$ to yield

$$\frac{1}{C} - \frac{1}{C_0} = kt$$

or

$$\frac{C}{C_0} = \frac{1}{1 + kC_0t} \tag{A2-2}$$

Since at $t = 0$ all the DNA is single-stranded, C_0 is equal to the total DNA concentration.

Equation A2-2 shows that the fraction of single-stranded DNA remaining in a reassociation reaction (C/C_0) is a function of C_0t, the product of the initial concentration and the elapsed time. Therefore the time course of a reassociation reaction is plotted conveniently as shown in Figure 2-17, to give a so-called "Cot" curve.

$C_0t_{\frac{1}{2}}$ is defined as the value of C_0t at which the reaction has proceeded to half completion ($C/C_0 = 1/2$). Substitution into Equation A2-2 shows that $C_0t_{\frac{1}{2}}$, which can be determined experimentally as shown in Figure 2-17, is the reciprocal of the second-order rate constant:

$$C_0t_{\frac{1}{2}} = \frac{1}{k}$$

A2-3 $C_0t_{\frac{1}{2}}$ is proportional to the sequence complexity of a DNA preparation

The *complexity* (X)* of a sheared DNA preparation is the length in nucleotide pairs of the longest nonrepeating sequence that could be produced by splicing together fragments in the population. For example, a preparation of repeating dAT copolymer (. .ATATAT. .) has a complexity of 2; a preparation of a repeating tetramer of the form $(ATGC)_n$ has a complexity of 4, and a preparation made from identical but internally nonrepeating DNA molecules 10^5 nucleotide pairs in length has a complexity of 10^5.

$C_0t_{\frac{1}{2}}$ is related directly to complexity. Consider the reassociation of two populations of denatured DNA fragments, one derived from identical nonrepeating molecules 10^5 nucleotide pairs in length such as the DNA of a bacteriophage ($X = 10^5$) and the other from identical nonrepeating molecules 5×10^6 nucleotide pairs in length such as the DNA of a bacterium ($X = 5 \times 10^6$). If the fragment size is 500 nucleotides, then in the first population the fragments representing any particular sequence and its complement make up 0.005% of the total DNA. In the second population, each 500-nucleotide sequence represents only 0.0001% of the total DNA. Therefore, if both populations are adjusted to the same *total DNA concentration*, the concentration of each 500-nucleotide sequence will be 50 times greater in the first population than in the second. Therefore $t_{\frac{1}{2}}$ for the first population will be 1/50 of $t_{\frac{1}{2}}$ for the second. Or, if the DNA concentrations are unequal, $C_0t_{\frac{1}{2}}$ for the first population will be 1/50 of $C_0t_{\frac{1}{2}}$ for the second.

The proportionality of $C_0t_{\frac{1}{2}}$ to complexity means that

$$X = KC_0t_{\frac{1}{2}} \tag{A2-3}$$

in which the proportionality constant K will depend on reaction conditions (cat-

---

*In the literature complexity often is denoted by C. In this book X is used to avoid confusion with the symbol C for DNA concentration.

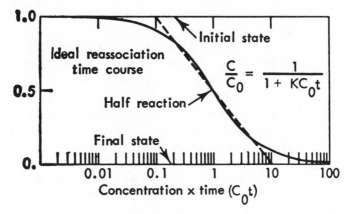

Figure 2-17. C_0t curve of an ideal second-order reassociation reaction. Note that 80% of the reaction occurs over a 2-log interval of C_0t. (From R. Britten and D. Kohne, *Science* **161,** 529, 1968.)

ion concentration, fragment size, etc.). Under the conditions generally employed (0.18N cation concentration, 400-nucleotide fragment size), $K \simeq 5 \times 10^5$ (liters × nucleotide pairs)/(moles of nucleotides × seconds), or

$$X = (5 \times 10^5) \; (C_0t_{\frac{1}{2}}) \tag{A2-4}$$

nucleotide pairs.

For organisms whose DNA contains no repeated sequences, X is simply the genome size (N) in nucleotide pairs, and N can be determined directly from measurements of $C_0t_{\frac{1}{2}}$ using Equation A2-3 or A2-4. All procaryotic genomes examined so far fall into this category. The reassociation of sheared DNA from these organisms goes to completion over about a 2-log interval of C_0t (see Figure 2-17), indicating that the concentration of all sequences in the population is the same.

A2-4 C_0t curves reveal the presence of repeated nucleotide sequences in eucaryotic genomes

Most eucaryotic DNA preparations reassociate over a C_0t range that is much greater than 2 logs, indicating that different sequences in the population are present at different frequencies, and therefore reassociate at different rates. Consider a hypothetical example of an organism whose genome consists of 50% repeated sequences of total complexity 10^5 nucleotide pairs, each repeated 10^4 times per genome (Component A), and 50% nonrepeated (unique) sequences of total complexity 10^9 nucleotide pairs (Component B). Therefore, N, the haploid genome size for this organism, is 2×10^9. If the DNA fragments representing these two components could be separated and reassociated independently, they would give the C_0t curves shown by dashed lines in Figure 2-18, with $C_0t_{\frac{1}{2}}$ values that differ by a factor of 10^4. When reassociated together, the two components behave *independently* to give the biphasic C_0t curve shown by the solid line in Figure 2-18. The following information can be obtained from such a curve:

(1) Genome fraction of each component (f). The fraction of the genome represented by each component can be determined by extrapolation from the end point of each component of the reassociation curve to the ordinate. In the example, $f_A = f_B = 0.5$.

(2) $C_0t_{\frac{1}{2}}$ (pure) for each component. The $C_0t_{\frac{1}{2}}$ value for each component in the mixture can be determined by extrapolating from the midpoint of each component of the reassociation curve to the abscissa. In the example, $C_0t_{\frac{1}{2}}$ (mixture) for A = 2×10^{-1} mole sec/liter and $C_0t_{\frac{1}{2}}$ (mixture) for B = 2×10^3 mole sec/liter. Note that these values are different from those determined from the C_0t curves of each component alone (Figure 2-18, dashed lines). This discrepancy arises because for the mixture, C_0 is taken to be the total DNA concentration, whereas for an isolated component, C_0 is taken to be the concentration of the component. To determine the complexity of a component, it is necessary to know the $C_0t_{\frac{1}{2}}$ value that the component would exhibit alone ($C_0t_{\frac{1}{2}}$ (pure)). This value can be obtained by multiplying the observed $C_0t_{\frac{1}{2}}$ (mixture) by the fractional contribution of the component to the genome:

$$C_0t_{\frac{1}{2}} \text{ (pure)} = fC_0t_{\frac{1}{2}} \text{ (mixture)} \tag{A2-5}$$

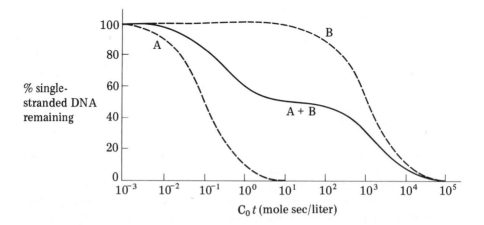

Figure 2-18. C_0t curves for whole DNA (solid line) and isolated frequency components (dashed lines) from a hypothetical eucaryotic organism.

In the example, the $C_0t_{\frac{1}{2} \text{ (pure)}}$ values for Components A and B are 10^{-1} and 10^3 mole sec/liter, respectively.

(3)   Repetition number (R). The *relative* frequencies of sequences in the various components of a mixture are inversely proportional to their $C_0t_{\frac{1}{2} \text{ (mixture)}}$ values. If the most slowly reassociating class is known to be unique (R = 1), or if the genome size, N, is known, then the absolute frequency or repetition number of the sequences in a component can be calculated. For any component, i, R_i is related to f_i, N, and X_i by the expression

$$R_i = \frac{f_i N}{X_i} \tag{A2-6}$$

This relationship also can be used to calculate the complexity X_i of a component if R_i, f_i, and N are known.

A2-5   Simple C_0t curves provide no information on arrangement of sequences or degree of complementary matching

(a)   The analysis described here provides no information on the lengths of unique and repeated sequences as functional units in the genome, or on the organization of these units. The question of organization is raised in Essential Concept 2-12, and an experimental approach to it is considered in Problem 2-29.

(b)   Complications in the analysis of C_0t curves arise when the members of a family of repeated DNA sequences are similar but not identical. The extent of mismatching in a component isolated after reassociation (using hydroxyapatite chromatography) can be determined from its absorbance-temperature profile in a thermal denaturation experiment (see WWBH Essential Concept 16-3). DNA duplexes that contain mismatches have a lower thermal stability than perfectly complementary duplexes of the same nucleotide composition. Generally, a drop of 1°C in the melting temperature (T_m) of a DNA duplex indicates about 1% mismatching. Some examples of mismatching and the complications that arise from it are considered in Problems 2-23 through 2-25.

REFERENCES

Where to begin

D. D. Brown, "The isolation of genes," *Scientific American* (August, 1973)
J. German, "Studying human chromosomes today," *Amer. Scientist* **58,** 182 (1970)

General

L. Stryer, *Biochemistry* Chapter 28 (Freeman, San Francisco, 1975)

J. D. Watson, *Molecular Biology of the Gene* Chapter 16 (W. A. Benjamin, Menlo Park, Calif., 1970, 2nd ed.)

H. L. K. Whitehouse, *Towards an Understanding of the Mechanism of Heredity* Chapters 17 and 18 (St. Martin's Press, New York, 1969, 2nd ed.)

Chromosome Structure and Function: Cold Spring Harbor Symposia on Quantitative Biology, Vol. 38 (Cold Spring Harbor Laboratory, Cold Spring Harbor, N. Y., 1973)

Chromosome structure

J. P. Baldwin, P. G. Boseley, E. M. Bradbury, and K. Ibel, "The subunit structure of the eucaryotic chromosome," *Nature* **253,** 245 (1975)

E. DuPraw, *DNA and Chromosomes* (Holt, Rinehart and Winston, New York, 1970)

T. C. Hsu, "Longitudinal differentiation of chromosomes," *Ann. Rev. Genetics* **7,** 153 (1973)

J. Huberman, "Structure of chromosome fibers and chromosomes," *Ann. Rev. Biochem.* **42,** 355 (1973)

R. Kevenoff, L. C. Klotz, and B. H. Zimm, "On the nature of chromosomal-sized DNA molecules," *Cold Spring Harbor Symposia on Quantitative Biology* **38,** 1 (1973)

R. O. Kornberg, "Chromatin structure: a repeating unit of histones and DNA," *Science* **184,** 868 (1974)

C. D. Laird, "DNA of *Drosophila* chromosomes," *Ann. Rev. Genetics* **7,** 177 (1973)

G. Stebbins, *Chromosomal Evolution in Higher Plants* (Addison-Wesley, London, 1971)

Chromosome organization

J. Bishop, "The gene numbers game," *Cell* **2,** 81 (1974)

J. Bishop and M. Rosbash, "Reiteration frequency of duck hemoglobin genes," *Nature New Biol.* **241,** 204 (1973)

H. G. Callan, "The organization of genetic units in chromosomes," *J. Cell Sci.* **2,** 1 (1967)

F. Crick, "General model for the chromosomes of higher organisms," *Nature* **234,** 25 (1971)

E. Davidson, B. Hough, C. Amenson, and R. Britten, "General interspersion of repetitive with nonrepetitive sequence elements in the DNA of *Xenopus*," *J. Mol. Biol.* **77,** 1 (1973)

E. Davidson and R. Britten, "Organization, transcription, and regulation in the animal genome," *Quart. Rev. Biol.* **48,** 565 (1973)

G. A. Galau, R. J. Britten, and E. Davidson, "A measurement of the sequence complexity of polysomal messenger RNA in sea urchin embryos," *Cell* **2,** 9 (1974)

J. Gall, E. Cohen, and D. Atherton, "The satellite DNAs of *Drosophila virilis," Cold Spring Harbor Symposia on Quantitative Biology* **38,** 417 (1973)

B. Judd, M. Shen, and T. Kaufman, "The anatomy and function of a segment of the X chromosome of *Drosophila melanogaster," Genetics* **71,** 139 (1972)

J. Morrow, S. Cohen, A. Chang, H. Boyer, H. Goodman, and R. Helling, "Replication and transcription of eucaryotic DNA in *Escherichia coli," Proc. Natl. Acad. Sci. (U. S.)* **71,** 1743 (1974)

S. Ohno, "Simplicity of mammalian regulatory systems," *Develop. Biol.* **27,** 131 (1972)

M. L. Pardue and J. Gall, "Chromosomal localization of mouse satellite DNA," *Science* **168,** 1356 (1970)

J. Paul, "General theory of chromosome structure and gene activation in eukaryotes," *Nature* **238,** 444 (1972)

C. Thomas, "The theory of the master gene," in F. O. Schmitt (Ed.), *The Neurosciences Second Study Program,* p. 973 (Rockefeller University Press, New York, 1970)

C. A. Thomas, Jr., "The genetic organization of chromosomes," *Ann. Rev. Genetics* **5,** 237 (1971)

P. Wensink, D. Finnegan, J. Donelson, and D. Hogness, "A system for mapping DNA sequences in the chromosomes of *Drosophila melanogaster," Cell,* in press, 1975

Appendix: Nucleic acid reassociation

R. J. Britten, D. E. Graham, and B. R. Neufeld, "Analysis of repeating DNA sequences by reassociation," in L. Grossman and K. Moldave (Ed.), *Methods in Enzymology,* Vol. 29, p. 363 (Academic Press, New York, 1974)

R. J. Britten and D. Kohne, "Repeated sequences in DNA," *Science* **161,** 529 (1968)

J. Wetmur and N. Davidson, "Kinetics of renaturation of DNA," *J. Mol. Biol.* **31,** 349 (1968)

PROBLEMS

2-1   Answer the following with true or false. If false explain why.
   (a)   The DNA of a eucaryotic chromosome is ordinarily one long double helix.
   (b)   Histones are extremely heterogeneous, acidic proteins.

(c)   Somatic cells passed on to progeny are important for perpetuation of the species.

(d)   The autosomal karyotype from different cells of an individual is invariant, but different individuals of the same species often have distinct autosomal karyotypes.

(e)   A chromatid is a chromosome that has been replicated but has not yet separated from its sister chromosome.

(f)   The centromere is always in the middle of a chromosome.

(g)   Most eucaryotes have more autosomes than sex chromosomes.

(h)   Certain stains permit the visualization of chromomeres in metaphase chromosomes.

(i)   The thickest dimension of the eucaryotic chromosomal fiber is probably about 100 Å.

(j)   Chromatin can be isolated from any tissue.

(k)   Proteins are synthesized in the puffs on giant chromosomes.

(l)   A polytene puff represents one or more activated genes.

(m)   The backbone of a lampbrush chromosome is a single DNA duplex.

(n)   Double-stranded DNA absorbs more UV light than single-stranded DNA.

(o)   Hydroxyapatite columns bind only DNA that is entirely double-stranded.

(p)   Some chromomeres are too small to encode structural genes.

(q)   Certain chromomeres contain as many as 10^5 nucleotide pairs, whereas others contain fewer than 3000 nucleotide pairs.

(r)   The mouse satellite DNA sequences at each centromere are tandemly linked.

(s)   Middle-repetitive sequences are made up of a single family of homologous sequences.

(t)   Most proteins are coded by unique DNA sequences.

(u)   The histone tetramer, believed to play an important role in chromosome structure, is comprised of the two most highly conserved histones.

(v)   Heterochromatic regions are visualized best on metaphase chromosomes.

(w)   Nucleated blood cells (e.g., lymphocytes) are ideal for the karyotype analysis of humans because they can be isolated easily from individuals. All that is necessary for the subsequent karyotype analysis is to treat the cells *in vitro* with colchicine and prepare appropriately stained metaphase plates for microscopic analysis.

(x)   A test for the presence of Barr bodies could be used to determine whether or not an alleged female should be allowed to compete as a female in the Olympic games.

2-2   (a)   The chromosomal abnormality that most commonly causes mongolism is called a trisomy because one particular autosome is present in _____ copies.

(b)   _____ are small, very basic proteins found in chromatin in a 1:1 (w:w) ratio with DNA and _____ .

(c)   In the mitotic cycle, sister _____ are held together at the _____ .

(d)   The average chromomere in polytene chromosomes contains about
_____ nucleotide pairs per chromatid.

(e)   The apparent unit of function in lampbrush as well as polytene
chromosomes is the _____.

(f)   Among the frequency classes of DNA sequences, the _____ class
generally constitutes the largest percentage in eucaryotic genomes.

(g)   The _____ technique permits labeled mRNA, synthesized *in
vitro,* to be hybridized with the DNA of metaphase or polytene chromosomes in
order to localize the position of repeated DNA sequences.

(h)   The control DNA sequences of Figure 2-16 probably would fall into the
_____ class of DNA.

(i)   The _____ model of chromosome structure predicts that most
structural genes should exist as families of identical, tandemly linked genes.

2-3   Only one human monosomy (a karyotype exhibiting a single copy of one
chromosome) has been described. Can you guess which human chromosome in
the monosomic state is compatible with life?

2-4   Klinefelter's syndrome (adult with no beard, testicles present but atrophied,
breasts developed) is found in humans with one Barr body, a Y chromosome, and
a normal autosomal karyotype. What is abnormal about the chromosomes of
these individuals?

2-5   To say that males are XY and females are XX does not explain how maleness or
femaleness is determined. What simple rules for sex determination in *Drosophila*
and man can you infer from the phenotypes given in Table 2-7?

Table 2-7. Phenotypes of various sex-chromosome combinations in *Drosophila* and man
(Problem 2-5).

Drosophila	Man
XO, sterile male	XO, sterile female
XY, normal male	XY, normal male
XX, normal female	XX, normal female
XYY, abnormal male (not viable)	XYY, normal (?) male
XXX, abnormal female (not viable)	XXY, abnormal male

2-6   Viable humans of normal appearance occasionally are found with 45 or even 44
chromosomes. How is this possible?

2-7   Certain women have two electrophoretic forms of the enzyme glucose-6-phos-
phate dehydrogenase in their red blood cells. When single-cell clones of cultured
fibroblasts are taken from such women, each clone shows only one of the two
electrophoretic patterns. Individual men exhibit either one pattern or the other,
but not both. What is a likely explanation for these observations?

2-8 How might you use the observation in Problem 2-7 to determine whether chronic myelocytic leukemia (a cancer of precursor red blood cells) in humans has a single-cell origin?

2-9 A single antibody-producing cell that contains genes for each of two variants of the same antibody can produce one variant or the other, but not both. Once such a cell has differentiated to express one or the other form of the gene, all its progeny cells synthesize the same antibody variant. However, in individual males as well as females who carry both genes, both variants of the antibody are found. What can you conclude from these observations about the chromosomal location of antibody genes?

2-10 Chromosomes of the mealy bug have some interesting properties. For example, when high doses of radiation are given to female mealy bugs, all of the offspring from crosses of these females with normal males die early in embryogenesis. In contrast, when the males are irradiated, the female offspring die, but the male offspring develop normally. Suggest an explanation for these observations. How would you submit this explanation to a further test?

2-11 In man the amount of DNA per diploid cell is about 6.4×10^9 nucleotide pairs.
(a) How long a double helix could be formed from this amount of DNA?
(b) Man has 23 pairs of chromosomes. What is the average length of DNA in each chromosome?
(c) The average length of human chromosomes at metaphase is about 6 μ. What average ratio of extended DNA length to condensed length (packing ratio) in the chromosome is necessary to accommodate the DNA molecules in these structures?
(d) Assume that man has 10^{14} cells. How long a double helix could be formed from the amount of DNA in a human individual?

2-12 A fly, *Chironomus,* has about 4.3×10^{-13} g of DNA per diploid cell and 3.4×10^{-9} g per polytene nucleus.
(a) How many DNA duplexes do polytene chromosomes of *Chironomus* contain, on the average?
(b) How many times must the diploid DNA be doubled to reach the polytene level?

2-13 *Drosophila* has a genome size of about 1.4×10^8 nucleotide pairs per haploid complement. There are about 5000 chromomeres in the corresponding polytene chromosomes. 95% of the DNA is in the bands and 5% is in the interband regions.
(a) How much DNA does the average band and the average interband region contain?
(b) What is the average number of gene-equivalents per band and per interband region, assuming an average gene size of 10^3 nucleotides?

2-14 The amino acid sequence of histone IV from calf thymus is given in Figure 2-19.
(a) What is unusual about the distribution of basic and hydrophobic amino acid residues in this molecule?

			5						10						15
1	Ser	Gly	Arg	Gly	Lys	Gly	Gly	Lys	Gly	Leu	Gly	Lys	Gly	Gly	Ala
16	Lys	Arg	His	Arg	Lys	Val	Leu	Arg	Asp	Asn	Ilu	Gln	Gly	Ilu	Thr
31	Lys	Pro	Ala	Ilu	Arg	Arg	Leu	Ala	Arg	Arg	Gly	Gly	Val	Lys	Arg
46	Ilu	Ser	Gly	Leu	Ilu	Tyr	Glu	Glu	Thr	Arg	Gly	Val	Leu	Lys	Val
61	Phe	Leu	Glu	Asn	Val	Ilu	Arg	Asp	Ala	Val	Thr	Tyr	Thr	Glu	His
76	Ala	Lys	Arg	Lys	Thr	Val	Thr	Ala	Met	Asp	Val	Val	Tyr	Ala	Leu
91	Lys	Arg	Gln	Gly	Arg	Thr	Leu	Tyr	Gly	Phe	Gly	Gly			

Figure 2-19.   The amino acid sequence of histone IV from calf thymus (Problem 2-14).

(b)   Can you suggest how these atypical distributions might be important for the function of histones?

2-15   Certain chromosomal regions remain heterochromatic throughout the lifetime of an organism (constitutive heterochromatin), whereas other chromosomal regions or entire chromosomes, such as the mammalian X, become heterochromatic only in particular stages or tissues (facultative heterochromatin). The satellite DNA of the mouse is constitutive heterochromatin.

(a)   What does this observation suggest about the function of mouse satellite DNA?

(b)   What might be an explanation for its localization to centromeric regions?

2-16   In *Drosophila* salivary gland cells with polytene chromosomes, the entire Y chromosome appears to be heterochromatic throughout the cell cycle. Males lacking the Y chromosome develop normally, but are sterile. How can you reconcile these two observations?

2-17   Which model in Figure 2-16 does each of the following observations support?

(a)   There are about 5000 chromomeres in *Drosophila* and about 10,000 in man.

(b)   In the nucleus there is a large amount of high-molecular-weight nuclear RNA, most of which does not reach the cytoplasm.

(c)   Judd and his co-workers have shown that 160 lethal or semilethal mutations in a portion of the *Drosophila* X chromosome fall into 12 complementation groups. There are 12 chromomeres in this same region of polytene X chromosomes.

(d)   The genes for hemoglobin and silk fibroin each are present only once in the haploid genome.

(e)   Occasionally, there are large differences in the amount of DNA per haploid set of chromosomes (this amount is designated the C-value) between closely related species. For example, *Ranunculaceae* (buttercups) show an 80-fold difference between two species that are similar in karyotype, but differ in size and morphology.

(f) Middle-repetitive fractions are composed of many different families of generally homologous, but not identical, sequences.

(g) At least some of the middle-repetitive sequences appear to be interspersed throughout the genome.

(h) Unique sequences constitute a large fraction of the genome of all eucaryotes.

2-18 Answer the following questions with reference to Figure 2-20.

(a) How many of these DNA preparations contain more than one frequency class of sequences? Explain your answer.

(b) If the genome size of $E.$ $coli$ is taken to be 4.5×10^6 nucleotide pairs, what is the genome size of T4?

(c) What is the complexity of mouse satellite DNA?

(d) Mouse satellite DNA represents 10% of the mouse genome. What is the repetition number for mouse satellite sequences, given that the haploid genome size is 3.2×10^9 nucleotide pairs?

(e) The calf genome is the same size as the mouse genome. What fraction of the calf genome is composed of unique sequences?

(f) The reassociation curve for $E.$ $coli$ DNA, shown in Figure 2-20, shows a $C_0t_{\frac{1}{2}}$ value of 8 mole sec/liter. How will this value change if $E.$ $coli$ DNA is allowed to reassociate in the presence of a 1000-fold excess of calf thymus DNA?

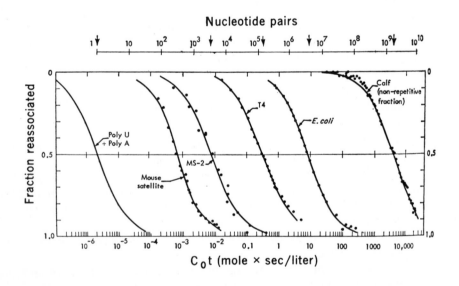

Figure 2-20. Reassociation of nucleic acids, sheared to 500-nucleotide fragments, from various sources (Problem 2-18). (From R. Britten and D. Kohne, $Science$ **161,** 529, 1968.)

2-19   The reassociation curves for calf-thymus DNA (O, Δ, and ●) and an internal standard, *E. coli* DNA (+), are shown in Figure 2-21.
(a)   How many frequency classes of DNA are present in the cow?
(b)   What fraction of the DNA does each class represent?
(c)   What is the $C_0t_{\frac{1}{2}\ (mixture)}$ of each class?
(d)   How many times are the most rapidly reassociating sequences repeated with respect to the most slowly reassociating sequences?
(e)   Given that the *E. coli* genome size is 4.5×10^6 nucleotide pairs, what is the complexity of the S fraction?

2-20   A small five-legged creature called a cloewg, which resembles the terrestrial cow, is brought back from an early morning expedition to Disneyland. The reassociation curve of its DNA is shown in Figure 2-22.
(a)   How many frequency classes of sequences are present in this DNA?
(b)   What fraction of the total genome does each class represent?
(c)   Explain how you might isolate the fastest (F) and the slowest (S) classes of DNA.
(d)   If you were to rerun C_0t curves on these isolated classes, what $C_0t_{\frac{1}{2}\ (pure)}$ values would you obtain?
(e)   What is the complexity (X) of the S class of DNA, assuming that the reassociation was carried out under standard conditions as defined in the Appendix to this chapter?

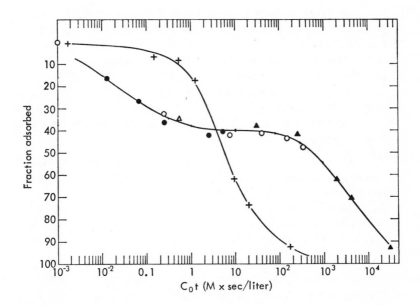

Figure 2-21.   Reassociation curves for calf-thymus and *E. coli* DNA's (Problem 2-19). (From R. Britten and D. Kohne, *Science* **161,** 529, 1968.)

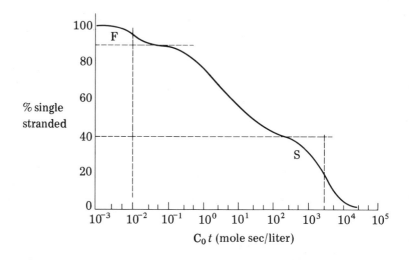

Figure 2-22. The reassociation curve of DNA fragments from the cloewg (Problem 2-20).

(f)   Given that the haploid genome size of the cloewg is 2.5×10^9 base pairs, how many copies of each sequence are present in the S class (i.e., what is the repetition number R)?

(g)   What is the repetition number for the F class?

(h)   What is the complexity of the F class?

(i)   How do the frequency classes of the Disneyland cloewg differ from those of the ordinary cow?

(j)   From the data presented, what can you determine about the organization of these sequences?

2-21   (a)   A C_0t curve for DNA from a Dipteran embryo is shown in Figure 2-23 (a). Would you expect a C_0t curve of polytene chromosomal DNA from this insect to be identical? Why?

(b)   Suppose that you obtain the C_0t curve shown in Figure 2-23(b) for polytene DNA. How does this C_0t curve differ from the preceding one, and how would you interpret this difference?

2-22   The most slowly reassociating fraction of the DNA from amphiuma, an amphibian, gives a $C_0t_{\frac{1}{2} \text{ (mixture)}}$ value of 1.6×10^5 mole sec/liter. Assuming that the reassociation was carried out under standard conditions and that R = 1 for the most slowly reassociating fraction, calculate the genome size of amphiuma.

(a)

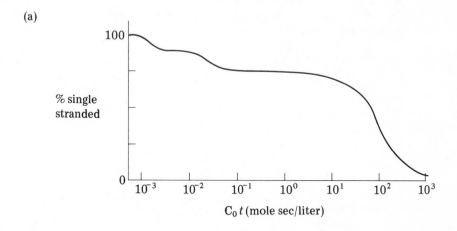

(b)

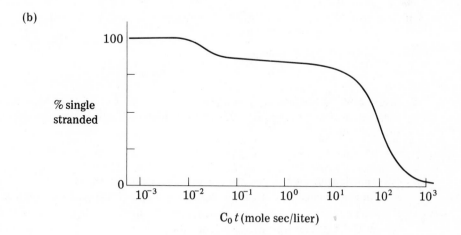

Figure 2-23. Hypothetical C_0t curves of DNA from a Dipteran fly. (a) Embryo DNA. (b) DNA from polytene nuclei (Problem 2-21).

2-23   Figure 2-24 shows absorbance-temperature profiles for various calf DNA's.
(a)   How would you explain the large difference in the melting profiles of the
reassociated repetitious and unique DNA's?
(b)   What is the significance of the difference between the profiles of native and
reassociated unique DNA?

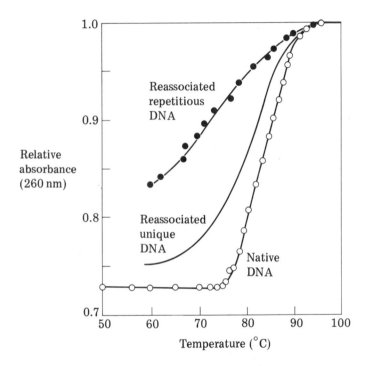

Figure 2-24.   Absorbance-temperature profiles of various calf DNA's. Native DNA is
high-molecular-weight (Problem 2-23). (Adapted from R. Britten and D. Kohne, *Science*
161, 529, 1968.)

2-24   Thermal denaturation experiments can yield valuable information about the
evolutionary relatedness of various frequency classes from different animals. In
one such experiment, a mixture of denatured ^{3}H-mouse DNA (290 μg/ml) and
^{14}C-rat DNA (0.086 μg/ml) are incubated together to a C_0t value sufficient for
reassociation of the repeated sequences but not the unique sequences. Due to the
difference in concentration of mouse and rat DNA there is essentially no self-
association of the rat sequences. The mixture then is passed over a
hydroxyapatite column. About 55% of the added DNA is bound to the column.
The bound material is eluted by passing a phosphate buffer through the column
while increasing its temperature linearly with time. The fractions then are
assayed for acid-precipitable radioactivity. The results are shown in Figure 2-25.

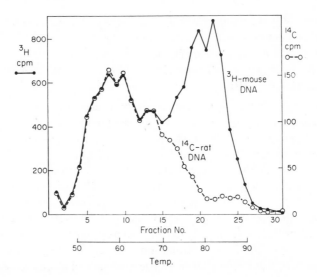

Figure 2-25. Thermal denaturation of reassociated repeated sequences of mouse and rat DNA (Problem 2-24). [From N. Rice in H. H. Smith (Ed.), *Evolution of Genetic Systems, Brookhaven Symposia in Biology,* **23,** Gordon and Breach, New York, 1972.]

(a)   Which column fractions correspond to the reassociated sequences with the least mismatch?

(b)   The repeated sequences of the mouse can be divided into two general categories, middle-repetitive and highly repetitive. Which column fractions correspond to each class?

(c)   Which of these frequency classes shows more homology with rat DNA?

(d)   The rat and mouse evolutionary lines diverged about 10^7 years ago. From this experiment, what can you say about the relative times of evolution of the middle- and highly repetitive sequences of DNA in the mouse?

2-25   Consider the data in Table 2-8.

Table 2-8. Effect of reassociation temperature on apparent percentage of repetitious DNA (Problem 2-25). (From D. Kohne, *Quart. Rev. Biophysics* **33**, 327, 1970.)

DNA	Temperature of reassociation (°C)	Percentage of genome as repetitious DNA
Human	51	43
	60	34
	66	25
Mouse	51	53
	60	34

(a)   Why does a change in the reassociation temperature affect the percentage of the genome classified as repetitious DNA?

(b)   What does this temperature effect indicate about mammalian DNA?

(c)   How would you expect the melting curves of DNA reassociated at 60°C and 51°C to differ?

2-26   Differential polyteny has been suggested to account for large differences in DNA content per nucleus within certain taxonomic groups that do not show corresponding differences in chromosome number. According to this hypothesis, the larger genomes evolved from the smaller ones by lateral duplication of the entire genome to produce multistranded chromosomes.

Vicia faba contains five times as much DNA per nucleus as Vicia sativa, but both these species of the pea genus have a haploid chromosome number of eight. Since the two species are closely related, it has been argued that the additional DNA of V. faba is unlikely to represent new genetic loci not present in V. sativa. Moreover, studies of chromosome structure using trypsin digestion have indicated that V. faba has more lateral strands in its chromosomes than does V. sativa. Thus V. sativa and V. faba would appear to be likely candidates for differential polyteny.

The differential polyteny hypothesis can be tested using DNA reassociation kinetics. An increase in polyteny increases the frequency of each sequence by the same amount and thus leaves the concentration of each sequence per gram of DNA unchanged. Since the rate of DNA reassociation is concentration dependent, differential polyteny will not change the rate of DNA reassociation. However, if new sequences are responsible for the increase in DNA content, the concentrations of individual sequences will be lower, and their rate of reassociation will decrease.

With these considerations in mind, the reassociation kinetics of V. faba and V. sativa DNA were studied with the results shown in Figure 2-26. Under the conditions used, sequences present once per haploid genome would be half reassociated at C_0t values of 3000 and 15,000 mole sec/liter for DNA from V. sativa and V. faba, respectively. Therefore Figure 2-26 represents the reassociation curves of repeated DNA only. Do the results support the differential polyteny hypothesis or not? Explain.

2-27   Restriction enzymes are endonucleases that cleave double-stranded DNA at particular nucleotide sequences four to eight residues in length (WWBH Chapter 16). It recently has been observed that a few lampbrush loops are digested to small fragments by a particular restriction enzyme, whereas most loops are unaffected. What are the implications of this observation?

2-28   (a)   In situ hybridization experiments with labeled 5S RNA easily can detect the clustered 5S DNA genes on Xenopus laevis chromosomes (see Table 2-5). The 5S RNA of Xenopus is about 150 nucleotides in length. Would it be possible to map a unique structural gene sequence on Drosophila polytene chromosomes with purified specific mRNA of the same specific radioactivity as the ^{3}H-labeled

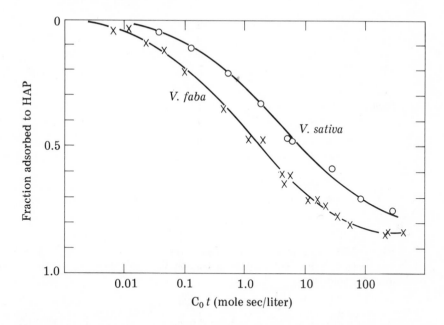

Figure 2-26. Reassociation kinetics of isolated repetitive DNA from *V. faba* and *V. sativa* monitored by adsorption to hydroxyapatite (Problem 2-26). (From N. Straus, *Carnegie Institution Year Book* **71,** 258, 1972.)

5S RNA used in the *Xenopus* experiments, assuming that the size of the structural gene mRNA is 1000 nucleotides, and that each structural gene is duplicated 1000-fold in *Drosophila* polytene chromosomes?

(b)   How would the experiment in (a) be affected if the unique mRNA contained a 1% contaminant consisting of mRNA for histones, labeled at the same specific radioactivity?

2-29   In an important series of experiments with *Xenopus laevis* DNA, hydroxyapatite chromatography was used to assay the reassociation of labeled fragments of various lengths with a 10,000-fold excess of unlabeled 450-nucleotide fragments. Figure 2-27 shows the fraction of labeled fragments reassociated sufficiently to bind to the column (R) as a function of labeled fragment size (L), at $C_0t = 50$ mole sec/liter. Under these conditions, only repetitive sequences associate.

In a separate experiment, larger quantities of 450-nucleotide and 1400-nucleotide fragments each were reassociated independently to $C_0t = 50$ mole sec/liter. The reassociated fractions then were isolated by hydroxyapatite chromatography and analyzed by thermal denaturation. The hyperchromicities observed were 17% and 10%, respectively, where hyperchromicity is defined as the increase in optical density from 60°C to 98°C expressed as a percentage of the 98°C value. The hyperchromicity of native *Xenopus* DNA is 27%.

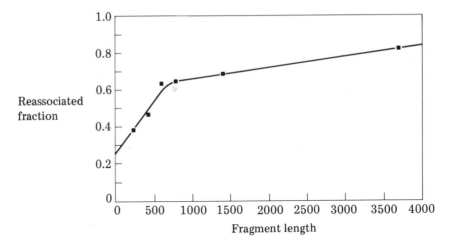

Figure 2-27. Reassociation of *Xenopus laevis* DNA at $C_0t = 50$ as a function of fragment length (Problem 2-29). (Adapted from E. Davidson and R. Britten, *Quart. Rev. Biol.* **48,** 565, 1973.)

(a) When *Xenopus* DNA is sheared to an average chain length of 4000 nucleotides, what fraction of the fragments consist entirely of unique sequences, based on the data in Figure 2-27?

(b) From the data in Figure 2-27, what fraction of the *Xenopus* genome is made up of repetitive sequences? Is your answer more, or less, reliable than the value for this fraction that would be obtained from a standard C_0t curve?

(c) Propose an interpretation for the change in slope of the curve in Figure 2-27 at L $\simeq$ 800 nucleotides, and the continued slow rise beyond this point.

(d) Based on the hyperchromicity experiment, what are the average lengths of repetitive sequence in 450-nucleotide and 1400-nucleotide fragments, respectively? How would you rationalize the differences in these lengths?

(e) Suggest an alternative and more direct experiment that would answer the first question in Part (d), using methods that you have encountered in this chapter.

ANSWERS

2-1 (a) True

(b) False. This description applies to nonhistone chromosomal proteins.

(c) False. This description applies to germ cells and the gametes they produce.

(d) False. The autosomal karyotype of a species is invariant apart from rare chromosomal abnormalities.

(e) True

(f)   False. The centromere can be at any point along the length of a chromosome.

(g)   True

(h)   False. The bands produced by the staining of metaphase chromosomes are not chromomeres. Chromomeres can be visualized only on polytene interphase chromosomes and the lampbrush chromosomes of meiosis.

(i)   True

(j)   True

(k)   False. Only RNA is synthesized in chromosome puffs.

(l)   True

(m)   False. A lampbrush chromosome contains two pairs of DNA duplexes (see Figure 2-12).

(n)   False. Double-stranded DNA absorbs less UV light.

(o)   False. Hydroxyapatite will bind molecules that are predominantly single-stranded if they include a small duplex region.

(p)   False. The average chromomere contains about 30 gene-equivalents of DNA. The smallest chromomeres contain three to five gene-equivalents.

(q)   True

(r)   True

(s)   False. Many different families exist in the middle-repetitive fraction.

(t)   True

(u)   True

(v)   False. Since all chromosomes are condensed during metaphase, heterochromatin and euchromatin cannot be distinguished by light microscopy.

(w)   False. Lymphocytes are not actively dividing cells, hence they cannot be trapped in metaphase with colchicine. Lymphocytes can be stimulated to divide by *in vitro* treatment with phytohemagglutin, a plant glycoprotein. These dividing cells then can be blocked in metaphase and subsequently analyzed for karyotype.

(x)   True

2-2   (a)   three

(b)   Histone, nonhistone chromosomal proteins

(c)   chromatids, centromere

(d)   30,000

(e)   chromomere

(f)   unique

(g)   *in situ* hybridization

(h)   middle-repetitive

(i)   master-slave

2-3   The X chromosome can be monosomic. This seems reasonable because the male normally has one X chromosome and the female has only one active X chromosome per cell. The XO condition, designated Turner's syndrome, produces short, sterile females. All autosomal monosomies apparently are incompatible with life.

2-4   These individuals are XXY. Barr-body formation does not occur unless there are two or more copies of the X chromosome present. For example, individuals with an XXXXY karyotype have three Barr bodies. All but one of the X chromosomes form Barr bodies, thereby maintaining the proper ratio of one active X per diploid set of autosomes.

2-5   In *Drosophila,* sex depends upon the ratio of X chromosomes to autosomes; a ratio of 1 or greater leads to a female. The Y chromosome is necessary for male fertility, but not for maleness. In humans, sex depends upon the presence or absence of the Y chromosome. Absence of Y leads to a female; one or more Y's leads to a male.

2-6   One or more small chromosomes can become attached to other chromosomes without the loss of any genetic material. Such an event is called a balanced translocation, since the resulting cells retain two copies of each autosome and the appropriate sex chromosomes.

2-7   The gene encoding glucose-6-phosphate dehydrogenase is on the X chromosome. In female mammals one of the two chromosomes is inactive in each cell. The inactivation is random; hence a woman who carries a gene for each of the two forms of the enzyme will have certain cells that express one variant and other cells that express the second. Males have a single X chromosome and, accordingly, can have only a single form of the enzyme.

2-8   If the cancer has a single-cell origin, then all leukemic cells will have the same enzyme variant in females who carry genes for both variants. If the cancer arises in many cells, then leukemic cells of both types should be found. In actual cases, leukemic cells from any one such female always contain only a single variant, suggesting that the cancer has a single-cell origin.

2-9   (1)   The antibody gene must be located on an autosome, since both variants are found in males who carry both genes.
(2)   The expression of only one of the two antibody genes in any single cell (allelic exclusion) suggests that parts of chromosomes can be inactivated in the same fashion as the female X chromosome in mammals.

2-10   The paternal set of chromosomes is inactive (heterochromatic) in male mealy bugs, whereas it must be active in females. The maternal set of chromosomes must be active in both males and females. One test of this explanation would be to expose tissues of mealy bugs to ^{3}H-uridine to measure chromosomal RNA synthesis. When this experiment is carried out on males, they demonstrate uptake of uridine over the euchromatic set (maternal), but not over the heterochromatic (paternal) set of chromosomes. Thus there are mechanisms in eucaryotes for selectively inactivating a part of one chromosome, a whole chromosome, or an entire haploid set of chromosomes.

2-11   (a)   The length of DNA helix occupied by one nucleotide pair is 3.4 Å. Therefore 6.4×10^9 nucleotide pairs $\times$ 3.4 Å $= 2.2 \times 10^{10}$ Å $= 220$ cm or 2.2 meters of double helix.

(b) 220 cm/46 = 4.8 cm per chromosome

(c) 4.8 cm/6 × 10^{-4} cm = 8.0 × 10^3

(d) 2.2 meters × 10^{14} = 2.2 × 10^{14} meters. (This length is 100 times the round-trip distance from the earth to the sun!)

2-12 (a) Polytene chromosomes arise by homologue pairing followed by replication of chromatids, to give the haploid number of chromosomes. The number of DNA strands per homologue in polytene nuclei is 3.4 × 10^{-9}/4.3 × 10^{-13} = 0.79 × 10^4. Since each polytene chromosome arose from a pair of homologues, it will contain twice this number of strands or 1.6 × 10^4.

(b) If n equals the number of doublings of the paired diploid chromosomes required to generate the polytene chromosomes, then $2^n = 0.79 × 10^4$ or $n = 13$.

2-13 (a) The average DNA contents are (1.4 × 10^8 × 0.95)/(5 × 10^3) = 3 × 10^4 nucleotide pairs per band and (1.4 × 10^8 × 0.05)/(5 × 10^3) = 0.1 × 10^4 nucleotide pairs per interband region.

(b) The averages are 30 gene-equivalents per band and one gene-equivalent per interband region. Thus either region has sufficient DNA for at least one structural gene.

2-14 (a) Seventeen of 27 basic residues are present in the N-terminal 45 residues, whereas only eight of 30 hydrophobic residues are present in this stretch. In contrast, 10 basic and 22 hydrophobic residues are present in the C-terminal 57 residues. Thus the N-terminal portion of the molecule is basic and the C-terminal portion is more hydrophobic. All histone molecules can be divided into a basic region as well as one with a more common amino acid composition.

(b) Probably the basic portions of histone molecules bind to DNA and the more hydrophobic portions interact with each other and other chromosomal components to promote the supercoiling that occurs in eucaryotic chromosomes. Histones may bind in the major groove of the DNA, since other molecules that bind to the minor groove can interact with chromatin as well as with free DNA. An elegant three-dimensional model has been proposed for histone IV, in which the first 18 residues form an α helix that fits perfectly into the major groove of the DNA helix, with five of six basic residues binding to one DNA strand and the sixth to the other strand. The large hydrophobic residue at position 10 fits perfectly into the major groove, leaving the appropriate sites for phosphorylation and acetylation properly exposed. The more hydrophobic C-terminal tail is free to interact with other hydrophobic components.

2-15 (a) DNA sequences in heterochromatic regions probably are not transcribed. Hence their function (if any) must be structural, for example, in matching of chromosomes at metaphase.

(b) These regions might bind the proteins of the spindle apparatus or be involved in the alignment of the centromeres of homologous chromosomes.

2-16 A chromosome or chromosomal region can be active for a single relatively short period during development. Thus the Y chromosome, although inactive in the specialized polytene nuclei of salivary gland cells, might be active at another stage. In fact, the Y chromosome develops a lampbrush configuration with at least five loops during meiotic prophase in diploid spermatocytes. All five loops must form to produce fertile sperm.

2-17 (a) This observation is compatible with all three models. However, the junk and master-slave models, which assume only one gene per chromomere, predict total gene numbers that seem improbably low to some biologists.

(b) This observation is most compatible with the control model. Most of the high-molecular-weight nuclear RNA could be composed of control sequences that are lost in the processing of messages during transport from the nucleus to the cytoplasm. A similar processing is observed for ribosomal RNA. It seems unlikely that junk DNA would be transcribed and it is not at all clear how the master-slave model would explain this observation.

(c) This observation is compatible with all three models.

(d) This observation is compatible with the junk and control models but is incompatible with the master-slave model.

(e) Superficially, this observation appears consistent with either the junk model or the master-slave model. However, in a similar comparison between two species of the fly *Chironomus,* certain bands were found to differ in DNA content by powers of 2, suggesting that single chromomeres can undergo one or more complete doublings in the course of speciation. No model has yet been proposed that can account for such geometric increases in DNA content of single bands.

(f) This finding is most consistent with the control model, assuming that the middle-repetitive sequences represent control elements. It could be rationalized in terms of the junk model, but is difficult to explain by the master-slave model.

(g) Same answer as (f).

(h) This observation is incompatable with the master-slave model.

2-18 (a) None of the preparations contains more than a single frequency class of sequences, since each shows about 80% reassociation over a 2-log interval of C_0t. If more than one frequency class were present, the C_0t curves would be broader.

(b) Genome size for procaryotes is equal to complexity, which is proportional to $C_0t_{\frac{1}{2}}$. From the curves in Figure 2-20 the $C_0t_{\frac{1}{2}}$ values for *E. coli* and T4 are 8 and 0.3, respectively. Therefore the genome size of T4 is $(4.5 \times 10^6)(0.3/8) = 1.7 \times 10^5$ nucleotide pairs.

(c) The $C_0t_{\frac{1}{2}}$ value for mouse satellite DNA is 7×10^{-4}. Therefore its complexity is $(4.5 \times 10^6)(7 \times 10^{-4}/8) = 400$ nucleotide pairs.

(d) Mouse satellite DNA comprises $(0.10)(3.2 \times 10^9) = 3.2 \times 10^8$ nucleotide pairs. If the complexity of the repeating sequence is 400 nucleotides, this sequence must be repeated 8×10^5 times.

(e) From Figure 2-20 the complexity of the calf unique sequence fraction is $(4.5 \times 10^6)(4 \times 10^3/8) = 2 \times 10^9$. Since these sequences are present only once, they comprise $2 \times 10^9/3.2 \times 10^9 = 60\%$ of the calf genome.

(f) The $C_0t_{\frac{1}{2}}$ value for *E. coli* DNA will not be affected significantly. The complementary sequences in a mixture appear to associate independently; their rate of association is dependent only on their concentration and independent of the concentration of noncomplementary sequences. This independence must be assumed in order to interpret any C_0t curve involving more than one frequency class of DNA sequences.

2-19 (a) Two frequency classes are present since there appear to be two second-order transitions in the reassociation curve. For convenience, these classes are designated the fast (F) and slow (S) classes.

(b) The fraction of the DNA in the F class is 0.4; the fraction in the S class is 0.6.

(c) $C_0t_{\frac{1}{2} \text{(mixture)(F)}} = 0.03$; $C_0t_{\frac{1}{2} \text{(mixture)(S)}} = 3000$.

(d) The concentration of sequences that reassociate rapidly is $C_0t_{\frac{1}{2}\text{(S)}}/C_0t_{\frac{1}{2}\text{(F)}}$ or 10^5 times the concentration of sequences that reassociate slowly. If the S fraction is made up of unique sequences, each of which occurs only once in the calf genome, the sequences of the F fraction must be repeated 10^5 times on the average.

(e) This question can be answered in either of two ways. Complexity is proportional to $C_0t_{\frac{1}{2}\text{(pure)}}$, which for the calf S fraction is $(3 \times 10^3) (0.6) = 1.8 \times 10^3$ mole sec/liter. From the curve shown in Figure 2-21, $C_0t_{\frac{1}{2}}$ for $E.\ coli = 5$. Therefore the complexity of the calf S fraction is $(4.5 \times 10^6) (1.8 \times 10^3/5) = 1.6 \times 10^9$ nucleotide pairs.

Alternatively, if you know that the haploid genome size of calf is 3.2×10^9 nucleotide pairs, and that the S fraction represents single-copy DNA, then the complexity of the S fraction must be equal to its portion of the genome, which is $(3.2 \times 10^9) (0.6) = 1.9 \times 10^9$ nucleotide pairs.

2-20 (a) There are at least four frequency classes. F and S appear to be single-frequency classes, but the intermediate fraction must contain more than one class, since it reassociates over a 3-log interval of C_0t.

(b) The fastest class represents 10%, the intermediate class 50%, and the slowest class 40% of the genome.

(c) The F class can be isolated by incubating the reassociating mixture to a C_0t of 10^{-1} mole sec/liter, adsorbing the double-stranded DNA to hydroxyapatite, and eluting with buffer containing a high concentration of salt. The S class may be isolated by incubating the mixture to a C_0t of 2.5×10^2 mole sec/liter, passing it through a hydroxyapatite column, and collecting the material that does not adsorb.

(d) In general, $C_0t_{\frac{1}{2}\text{(pure)}} = fC_0t_{\frac{1}{2}\text{(mixture)}}$. Therefore, $C_0t_{\frac{1}{2}\text{(pure)(F)}} = 0.1 \times 10^{-2} = 10^{-3}$ mole sec/liter, and $C_0t_{\frac{1}{2}\text{(pure) (S)}} = 0.4 \times 0.25 \times 10^4 = 10^3$ mole sec/liter.

(e) Under standard experimental conditions, complexity is related to $C_0t_{\frac{1}{2}\text{(pure)}}$ by the proportionality constant 5×10^5 (liters $\times$ nucleotide pairs)/(seconds $\times$ moles of nucleotide). Hence, for the S class, whose $C_0t_{\frac{1}{2}\text{(pure)}}$ is 10^3 mole sec/liter, $X = 5 \times 10^8$ nucleotide pairs.

(f) The S class represents 40% of the total genome, or 1×10^9 nucleotide pairs. The complexity of the S class is half this value; hence there must be two copies of every sequence per genome, that is, R = 2.

(g)   The ratio of R values for different frequency classes in the population can simply be read off of the original C_0t curve as the ratio of $C_0t_{\frac{1}{2}}$ values. For the S and F classes, this ratio is $2.5 \times 10^3/10^{-2} = 2.5 \times 10^5$. Since the repetition number for the S class is 2, the repetition number for the F class is 5×10^5.

(h)   The complexity of this class can be obtained as in Part (e) from the $C_0t_{\frac{1}{2}(pure)}$ for the F class (10^{-3} mole sec/liter). The complexity of the F class is $(5 \times 10^5)(10^{-3}) = 500$ nucleotide pairs.

(i)   In marked contrast to all earthly creatures, the genome of the cloewg contains no unique sequences. Forty percent of its genome is composed of sequences repeated only twice, and ten percent is composed of short sequences of complexity 500 that are present in half a million copies. The remainder of the genome consists of two or more classes of DNA with intermediate repetitiveness.

(j)   This analysis tells you nothing about the arrangement of sequences within the genome. For example, the F class might represent 5×10^5 copies of a sequence of 500 contiguous nucleotide pairs, or 5×10^5 copies each of 5 different sequences of 100 contiguous nucleotide pairs. The repeated sequences could be in tandem, or interspersed between nonrepeating sequences. Other kinds of experiments are required to distinguish among these possibilities.

2-21   (a)   The curve in Figure 2-23 (a) has about 10% highly repetitive, 15% middle-repetitive, and 75% unique. A C_0t curve for polytene DNA should be identical to that of embryo DNA only if the C_0 for each frequency class does not change upon polytenization.

(b)   The polytene C_0t curve suggests that highly repetitive DNA is lost upon polytenization. This would suggest that the highly repetitive (centromeric) DNA present in most chromosomes has not been amplified. *Drosophila*, in fact, shows this behavior; its four polytene chromosomes are closely associated at their centromeric regions in a highly heterochromatic structure called the chromocenter.

2-22   Under standard conditions, $X = (C_0t_{\frac{1}{2}(pure)})(5 \times 10^5)$. $C_0t_{\frac{1}{2}(pure)} = fC_0t_{\frac{1}{2}(mixture)}$. Therefore $X = (fC_0t_{\frac{1}{2}(mixture)})(5 \times 10^5)$. X is related to N by the expression $X = fN/R$. Equating the two expressions for X gives $N = (C_0t_{\frac{1}{2}(mixture)})(5 \times 10^5) = (1.6 \times 10^5)(5 \times 10^5) = 8 \times 10^{10}$ nucleotide pairs.

2-23   (a)   The smaller relative absorbance change, shallower slope, and lower T_m of the rapidly reassociating DNA indicates the presence of single-stranded regions and considerable mismatching in the reassociated sequences.

(b)   The unique fraction would be expected to reassociate almost perfectly if each sequence finds its complement. Thus the difference between these profiles probably is due in part to the small fragment size of the reassociated DNA rather than to mismatching.

2-24   (a)   The sequences with the least mismatch melt at the highest temperature. Therefore fractions 22 to 27 represent the least mismatched sequences.

(b)   Since the highly repetitive mouse sequences (satellite DNA) are nearly identical ($\leq 6\%$ mismatch), the reassociated duplexes will have a high melting temperature. The less similar middle-repetitive sequences form duplexes with more mismatches that melt at a lower temperature. Therefore fractions 5-12 are middle-repetitive sequences, and fractions 16-26 are highly repetitive sequences.

(c)   The middle-repetitive sequences show more homology.

(d)   The satellite DNA must have evolved in the mouse *after* its divergence from the rat evolutionary line. Thus the appearance of the mouse satellite DNA is a recent evolutionary event. In contrast, the middle-repetitive families apparently existed prior to divergence. Other satellites also appear to have evolved in recent times. Possible mechanisms for the evolution of satellite DNA are discussed in Chapter 7.

2-25   (a)   The fraction of DNA that reassociates rapidly increases as the temperature is lowered because sequences with more mismatches become able to form stable duplexes.

(b)   This effect indicates that mammalian repetitious DNA contains related sequences that have varying degrees of homology.

(c)   The 60°C temperature imposes a higher *criterion* for reassociation; only well-matched duplexes are stable enough to form at this temperature. At 51°C, more mismatching is permitted. Therefore duplexes formed at 60°C will give melting curves with greater total increase in absorbance, higher melting point, and steeper slope than those of duplexes formed at 51°C.

2-26   The reassociation curves show that the average renaturation rate of DNA from *V. faba* is *faster* than that of *V. sativa*. This result suggests that if the genome of *V. faba* grew from a genome very much like that of *V. sativa*, it did so by multiplying some sequences more than others. Therefore the results rule out the hypothesis that *V. faba* is a simple polytenic derivative of the present-day genome of *V. sativa*.

2-27   The affected loops must have several copies of the particular nucleotide sequence that the restriction enzyme recognizes; accordingly, these loops are cleaved into small fragments. This result implies a tandem repetition of similar or identical sequences.

2-28   (a)   The 24,000 repeats of the 5S DNA sequence in *Xenopus laevis* are divided among most, if not all, of the 18 chromosomes in the haploid genome (Table 2-5). Accordingly, each chromosome has, on the average, 1333 copies of the DNA sequence complementary to 5S RNA, which is 150 nucleotide pairs in length. Thus the average total size of the 5S DNA clusters that associate with labeled 5S RNA in a hybridization experiment is 2×10^5 nucleotide pairs. A unique gene 1000 base pairs in length that is amplified 1000-fold in the polytene chromosome represents 10^6 nucleotide pairs that can associate with labeled mRNA. Therefore it should be possible to map amplified unique sequences on polytene chromosomes by *in situ* hybridization.

(b)   There are probably about 100 copies of each of the histone genes in the normal haploid chromosomal complement of *Drosophila* [Essential Concept 2-15(b)]. Hence in polytene chromosomes there would be 100 times as many copies of each histone gene as there would be copies of each unique gene. Since the histone genes are roughly half the size of the average unique gene, each class of histone genes will become labeled with 50 times as much radioactivity as the unique gene, because the *in situ* hybridization experiments are carried out with excess RNA. Contamination with mRNA copies of repeated DNA sequences can be a serious limitation in attempts to map unique sequences by *in situ* hybridization.

2-29   (a)   At $C_0t = 50$, labeled fragments that contain both repetitive and unique sequences will reassociate partially and be retained by the column, but fragments that consist entirely of unique sequences will not. The value of $R = 0.8$ at a fragment length of 4000 nucleotides indicates that 80% of these fragments contain at least one repetitive sequence. The remaining 20% must consist entirely of unique sequences.

(b)   As the length of the labeled fragments decreases, the probability becomes higher that each fragment consists entirely of unique or entirely of repetitive sequences. At $C_0t = 50$, only repetitive sequences will reassociate. Therefore, extrapolation of the curve in Figure 2-27 to $L = 0$ gives a value for the fraction of the genome that consists of repetitive sequences. This value of 0.25 is more reliable than that obtained from a standard C_0t curve analysis, in which the fragments are generally 400 to 500 nucleotides in length and have a higher probability of including both unique and repetitive sequences.

(c)   The change in slope suggests that in 60% of the genome, there is an average of one repetitive sequence and one longer unique sequence per 800 nucleotides, and that in the remaining 40% of the genome repetitive sequences are much less frequent. At $L = 800$ almost all of the fragments from the 60% portion of the genome contain at least one repetitive sequence. Further increases in L beyond 800 simply produce fragments from this portion of the genome with two or more repetitive sequences, and therefore do not increase the fraction of these fragments that can partially reassociate. However, as L increases up to 4000, more of the fragments from the remainder of the genome contain a repetitive sequence, so that the curve continues to rise gradually.

(d)   Native *Xenopus* DNA, which is 100% duplex, gives a hyperchromicity of 27%. The hyperchromicity of 17% for reassociated 450-nucleotide fragments indicates that the population is 17/27 or 63% duplex. The duplex DNA represents repetitive sequence; therefore the average amount of repetitive sequence per fragment is $0.63 \times 450 = 280$ nucleotides. A similar calculation for the 1400-nucleotide fragments yields a value of 530 nucleotides of repetitive DNA per fragment. The latter value is greater because at $L = 1400$ most fragments contain more than one repetitive sequence.

(e)   Each of the fragments isolated in the hyperchromicity experiment will contain duplex repetitive DNA and single-stranded unique DNA. If the population of fragments is digested with a single-strand-specific nuclease, such as the S_1 nuclease from *Aspergillus* (Appendix, Section A2-1), only the duplex regions of each fragment will remain. The size distribution of these regions then can be estimated directly by zone sedimentation, gel filtration, or polyacrylamide gel electrophoresis.

3 Chromosome Replication And Segregation

Eucaryotes have evolved elaborate mechanisms for maintaining the normal chromosome number during somatic cell division, and for reassorting and recombining the chromosomes of two organisms by sexual reproduction. In the process of cell division the chromosomes replicate, condense, and segregate to the two daughter cells by either of two processes, mitosis or meiosis. Mitosis preserves the chromosome number in divisions of somatic cells. Meiosis reduces the diploid chromosome number to the haploid number in the production of sperm and egg cells. Recombination between homologous chromosomes occurs during meiosis. This chapter considers diploidy and the sexual cycle, chromosomal DNA replication, the mechanics of mitosis and meiosis, and the genetic consequences of chromosome reassortment and recombination during meiosis.

ESSENTIAL CONCEPTS

3-1   Diploid cells have two copies of each autosomal gene

(a)   In the somatic cells of a diploid individual, there are two copies of every autosome, one inherited from each of the individual's parents, and hence two copies of every autosomal gene. Different forms of the same gene are called *alleles*. For example, β^A (normal) and β^S (sickle-cell) are alleles of the gene for the β chain of hemoglobin A in humans. Individuals who carry two different alleles of a gene (e.g., β^A/β^S) are said to be *heterozygous* for that gene, whereas individuals with two copies of the same allele are said to be *homozygous*. When only one copy of a gene is present, for example, on the X chromosome of a human male, the individual is said to be *hemizygous* for that gene.

(b)   When one allele exerts its genetic effect regardless of the presence or absence of a second allele, the first is said to be *dominant* and the second *recessive*. Because recessive genes often show no effects in the heterozygous condition, the expressed traits or *phenotype* of an individual organism does not always reflect its genetic constitution or *genotype*. For example, β^A in humans is dominant and β^S is recessive. Therefore, individuals who are homozygotes of genotype β^S/β^S show the mutant phenotype of sickle-cell anemia, whereas β^A/β^S heterozygotes are nearly indistinguishable in their physiological phenotype from β^A/β^A homozygotes.

Some alleles are expressed to varying degrees in different individuals.

Among individuals carrying such an allele, the fraction that shows the corresponding phenotype trait is defined as the *penetrance* of the allele.

3-2   Diploid (2n) somatic cells alternate with haploid (n) gametes at each generation in organisms that reproduce sexually

(a)   The gametes of higher eucaryotes are haploid, normally containing one complete set of autosomes and a sex chromosome. The life of a new individual begins with the process of *fertilization,* in which a male gamete (sperm) fuses with a female gamete (egg) to form a diploid embryo or *zygote*. In humans and

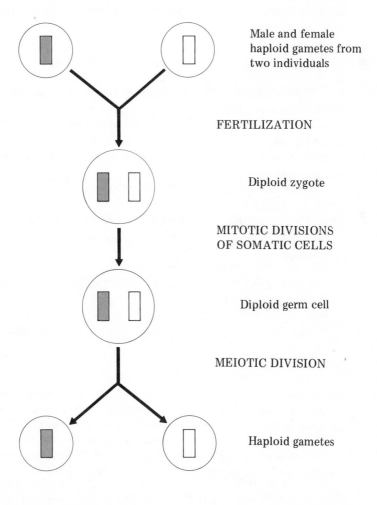

Male and female
haploid gametes from
two individuals

FERTILIZATION

Diploid zygote

MITOTIC DIVISIONS
OF SOMATIC CELLS

Diploid germ cell

MEIOTIC DIVISION

Haploid gametes

Figure 3-1.   The sexual cycle in diploid eucaryotes. The dark and light rectangles represent a pair of homologous chromosomes.

Drosophila the egg always carries an X chromosome; therefore the sperm, which can carry either an X or a Y chromosome, determines the sex of the embryo.

(b)   The zygote gives rise to all the somatic cells of the individual by a series of mitotic divisions, in which the chromosome complement is maintained at 2n. The average number of somatic-cell doublings that occurs during development is given by $\log_2 N$, in which N is the number of somatic cells in the adult organism. Thus the average number of doublings varies widely; it is about 10 in the 600-celled soil nematode *Caenorhabditis* and about 50 in a 10^{14}-celled human.

(c)   Specialized *germ cells* in the adult organism undergo reduction division or meiosis to produce haploid (n) gametes, which can undergo fusion to begin a new generation. The alternation between haploid and diploid chromosome number at each generation is called the *sexual cycle*. It is diagramed schematically in Figure 3-1.

3-3   Chromosomes replicate before each mitotic division

(a)   Chromosomal DNA replicates during interphase of the cell cycle when the chromosomes are in the form of extended chromatin fibers. Chromatin is replicated as a series of tandemly joined replication units, each representing about 30μ of extended DNA. Replication begins at the center of each unit and proceeds in both directions to the unit boundaries, where the growing forks from two adjacent units meet [Figure 3-2(a)].

   Different units along a chromatin fiber can begin replication at different times. In general, the DNA of heterochromatic regions replicates late in interphase after most of the euchromatic DNA has replicated. The spacing and differential initiation of adjacent replication units can be seen in preparations of partially replicated chromosomal DNA, which exhibits patterns of loops interspersed with nonreplicated segments as shown in Figure 3-2(b).

(b)   Based on the limited evidence so far available, the molecular events of replication at each growing fork are similar to those observed in procaryotes (WWBH Chapter 17). Pulse-chase radioactive labeling experiments show that chain growth occurs by synthesis of short fragments that are joined subsequently

Figure 3-2.   (a) Model for replication of a chromosomal DNA segment. The four ▶ diagrams represent progressive stages in the duplication of two adjacent replication units. Solid lines and dashed lines indicate parental and newly synthesized DNA, respectively. Letters indicate positions of origins (o) and termini (t). (1) Prior to initiation of replication. (2) Replication begun in right unit. (3) Replication begun in left unit, completed at termini of right unit. (4) Replication completed in both units. (Adapted from J. A. Huberman and A. D. Riggs, *J. Mol. Biol.* **32,** 327, 1968.) (b) Electron micrograph of replicating chromosomal DNA showing loops resulting from pairs of replication forks. Inset shows a tracing of a single continuous replicating duplex. (From H. J. Kriegstein and D. S. Hogness, *Proc. Natl. Acad. Sci. (U. S.)* **71,** 135, 1974.)

(a)

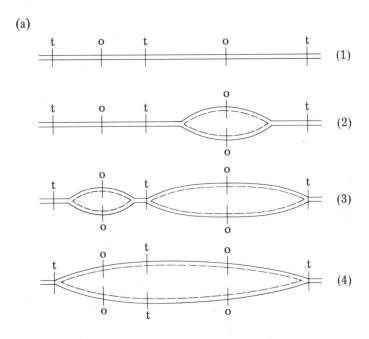

(b)

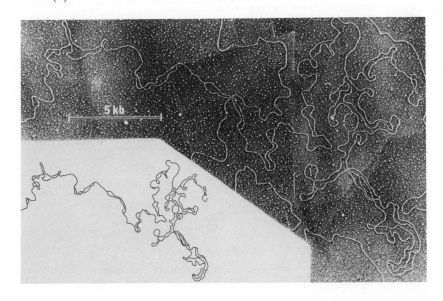

to longer daughter strands. Growing points progress along the DNA duplex at about 2.5μ/min. In contrast to procaryotic DNA, which replicates in association with cell membrane, chromosomal DNA appears to replicate throughout the nucleus.

(c)   The final product of replication is a duplicated interphase chromosome consisting of two identical chromatin fibers or sister chromatids, attached together in the region that will become the centromere during the subsequent metaphase. Little is known about how histone and nonhistone proteins are added to the fibers following DNA replication.

3-4   Chromosome segregation is controlled by the spindle apparatus

The *spindle apparatus* (Figure 3-3) controls the segregation of sister chromatids during mitosis, and of homologous chromosomes during meiosis. The spindle appears to be organized by a pair of *centrioles,* one at each pole of the cell. The spindle fibers are *microtubules,* formed by polymerization of the protein *tubulin.* At the onset of mitosis, microtubules grow radially from each centriole as the nuclear membrane disintegrates. Tubules from each centriole attach to each chromosome at the centromere, and somehow align the chromosomes in a planar configuration at the cell equator so that sister chromatids are connected to opposite poles by bundles of microtubules. Other microtubules do not attach to chromosomes, but appear to form continuous bridges between the two poles. Then, in response to some unknown signal, the centromeres separate, and the microtubules connected to them pull sister chromatids apart toward opposite poles of the cell. The necessary force probably is generated by cross-bridges between these microtubules and those that are not attached to chromosomes. The two kinds of tubules are thought to slide relative to each other by a mechanism analogous to the sliding of actomyosin filaments in muscle.

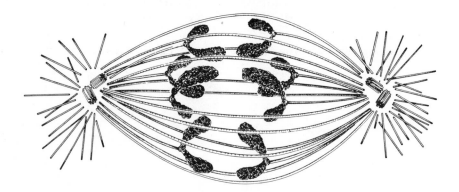

Figure 3-3.   The spindle apparatus at mitotic metaphase. (From E. J. DuPraw, *DNA and Chromosomes,* Holt, Rinehart, and Winston, New York, 1970.)

3-5   **Mitosis provides for segregation of sister chromatids to daughter cells in somatic cell divisions**

Mitosis commonly is described in four stages, as diagramed in Figure 3-4.

(1)   During *prophase,* the first stage of mitosis, chromatin fibers condense, and individual chromosomes become visible in the light microscope as pairs of chromatids connected at a centromere. Prophase ends with the breakdown of the nuclear membrane and the formation of the spindle, which begins to orient the chromosomes.

(2)   In *metaphase* the chromosomes become aligned at the cell equator by the spindle apparatus, as described in the preceding section.

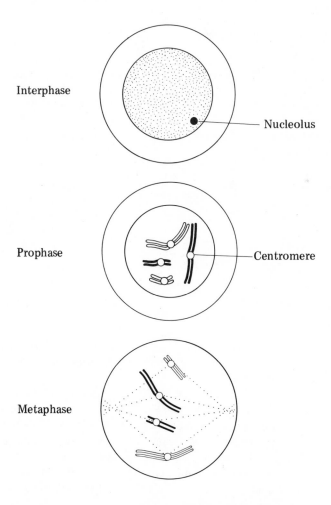

Figure 3-4.   The stages commonly distinguished in describing mitosis.

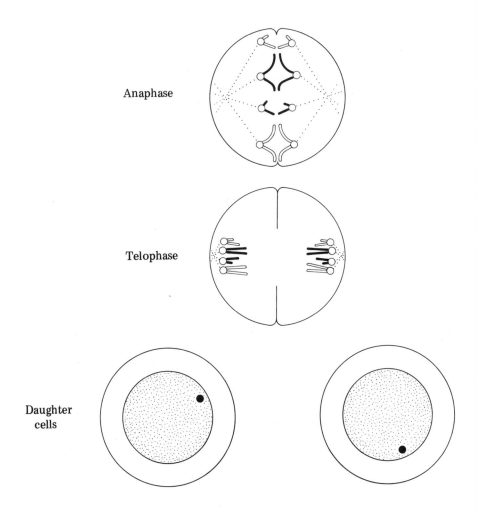

Anaphase

Telophase

Daughter
cells

Figure 3-4. (continued)

(3)   At the onset of *anaphase* each pair of sister chromatids is pulled apart
      to form two sister chromosomes, which move to opposite poles of
      the spindle.

(4)   In the final stage, *telophase,* the microtubules of the spindle
      apparatus depolymerize and nuclear membranes re-form around the
      condensed chromosomes at each pole. At about the same time, the
      plasma membrane is pinched together between the poles by an equa-
      torial band of contractile *microfilaments,* to cause actual cell sepa-
      ration or *cytokinesis*. During this stage the chromosomes become ex-
      tended again, and the two daughter cells return to the interphase
      period of the cell cycle.

3-6 Meiosis forms haploid gametes with new combinations of genes

(a) Meiosis is the specialized division process by which diploid germ cells give rise to haploid gametes. Meiosis involves one chromosomal replication and two cell divisions, with the net effect of reducing the chromosome number by a factor of two in each of four resulting granddaughter cells.

(b) Two features of meiosis are of key importance to the process of heredity in eucaryotes (see Essential Concept 3-8).

> (1) Each diploid germ cell contains two homologues of each chromosome, one inherited from the individual's maternal parent and one from its paternal parent. During the first meiotic division, these homologues become paired and aligned at the cell equator with the maternal homologue of each pair oriented toward either pole with equal probability. Thus when the homologues are separated by the spindle apparatus, each pole, and each subsequently resulting pair of gametes, receives a random mixture of maternal and paternal homologues.

> (2) While chromosomes are paired during the first meiotic prophase, segments of DNA are physically exchanged between maternal and paternal homologues by crossing-over or *genetic recombination*. This process can yield new combinations of alleles on individual chromosomes, which subsequently are incorporated into gametes and passed on to the next generation.

(c) The stages commonly distinguished in describing meiosis are diagramed in Figure 3-5.

> (1) From the genetic point of view the most important stage is *Prophase I*, which can be divided into five substages. The chromosomal DNA has replicated prior to the first of these substages, called *leptotene*, in which chromosomes first become visible. During *zygotene*, homologous chromosomes undergo pairing or *synapsis*. At this stage the two sister chromatids of each homologue are not yet visible. In the next substage, *pachytene*, sister chromatids become visible, and the resulting *tetrads* begin to form regions of contact and crossing-over between homologous nonsister chromatids. During *diplotene* and *diakinesis*, the next substages, regions of crossing-over (*chiasmata*) become more visible as the pairs of sister chromatids in each tetrad pull away from each other over most of their length.

> (2) In *Metaphase I* and *Anaphase I*, the tetrads become aligned at the cell equator and separated by the spindle apparatus into homologues that migrate to opposite poles. Re-formation of the nuclear membrane around the chromosomes at each pole during Telophase I completes the first meiotic division.

> Note that the stages so far described involve only the pairing of homologous chromosomes and their subsequent separation. The net

First Meiotic Division
  Prophase I
    Leptotene——chromosomes become visible
               (each containing two DNA
               double helices).

    Zygotene—— homologous chromosomes pair.

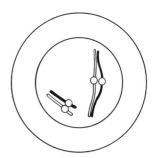

    Pachytene——chromosomes split into visible
              sister chromatids to form tetrads
              and crossing over begins.

    Diplotene—— chiasmata become visible.

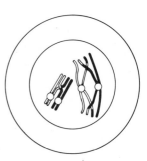

Figure 3-5. Stages commonly distinguished in describing meiosis. Prophase I, the most important for genetic analysis, is divided arbitrarily into five substages. The stages of the second meiotic division parallel those of mitosis except that there is no chromosomal doubling prior to Prophase II.

Diakinesis—chromosome arms rotate relative to one another so that chiasmata assume their definitive cross shapes.

Metaphase I — tetrads orient on the first division spindle.

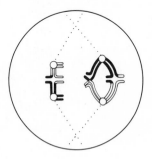

Anaphase I — centromeres of homologous chromosomes separate.

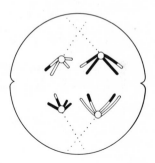

Telophase I — two daughter nuclei form.

Second Meiotic Division
  Prophase II — chromosomes condense.

  Metaphase II—chromosomes orient on the equators of the second division spindles.

  Anaphase II — centromeres divide and separate.

  Telophase II—daughter nuclei form.

Figure 3-5. (continued)

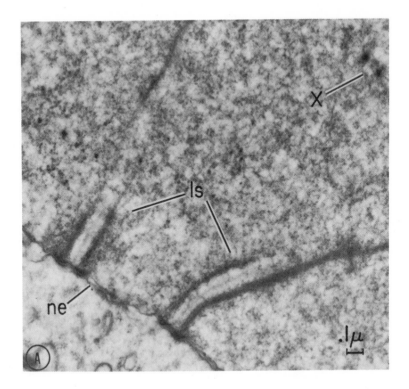

effect of this process is to allow homologous nonsister chromatids to exchange genetic material, and then to separate the two parental homologues of each chromosome into different daughter cells.

(3) The second meiotic division is essentially a mitosis, which takes place without prior chromosome replication. The sister chromatid pairs in each haploid set again condense, align, and separate into daughter nuclei, each of which thereby acquires a haploid set of chromosomes.

(4) In spermatogenesis, each of the four daughter cells from meiosis forms a functional male gamete or sperm. In the oogenesis of higher animals, only one of the four daughter cells becomes a functional female gamete or egg; the other three remain as small abortive cells with little cytoplasm, called polar bodies.

3-7 Recombination of chromosomal DNA occurs during the first stage of meiosis

(a) Synapsis of chromosomes in meiotic Prophase I brings the DNA molecules of homologous nonsister chromatids into intimate contact through structures called *synaptinemal complexes*. Composed of histonelike proteins with a width of about 2000 Å, these complexes mediate pairing over the entire length of each

◀ **Figure 3-6.**   Electron micrograph of a thin section of a rat germ cell (spermatocyte) nucleus showing several synaptinemal complexes and their attachment to the nuclear envelope. Each complex consists of two lateral elements containing the homologues and one central element; the complexes fade out after a short distance because they do not lie entirely in the plane of the section. The meaning of the symbols is as follows: ne = nuclear envelope; ls = longitudinal section of a synaptinemal complex; X = cross section of a synaptinemal complex. (From M. J. Moses and J. R. Coleman, "Structural patterns and the functional organization of chromosomes," in M. Loeke (Ed.), *The Role of Chromosomes in Development, 23rd Symp. Soc. Stud. Growth and Develop.* 1964.)

pair of homologues during tetrad formation, and terminate on the nuclear membrane at both ends (Figure 3-6). The complexes also may play an active role in recombination of the chromosomal DNA. Synaptinemal complexes disperse following the pachytene phase of Prophase I.

(b)   Recombination results from exchanges of individual DNA strands between the duplex DNA molecules of synapsed homologous nonsister chromatids. Recombinant chromosomes that form during Prophase I appear in separate gametes, following separation of the two homologues in the first meiotic division and separation of the sister chromatids of each homologue in the second (Figure 3-7). The proposed molecular mechanism of recombination illustrated in Figure 3-8 is supported by genetic evidence and the limited biochemical evidence so far available.

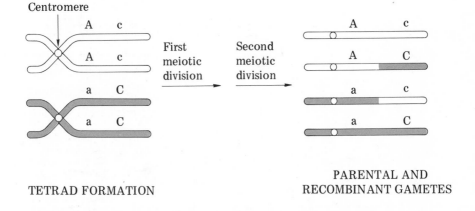

TETRAD FORMATION

PARENTAL AND
RECOMBINANT GAMETES

Figure 3-7.   Consequences of recombination between nonsister chromatids on homologous chromosomes. During the tetrad stage, an average of two or three crossover events occurs between each pair of sister chromatids. The homologous pairs separate during the first meiotic division and the sister chromatids separate at their centromeres during the second meiotic division, so that each gamete receives one chromatid from the tetrad stage.

(a)

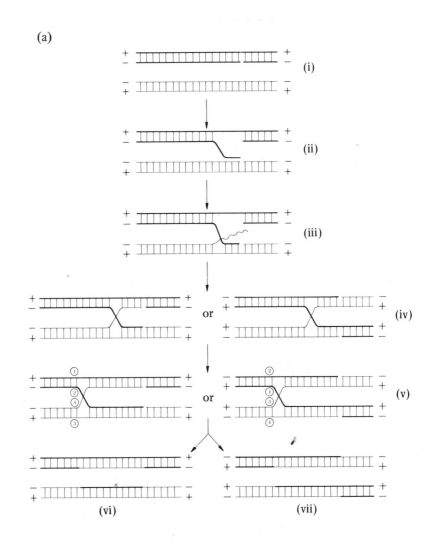

(b)

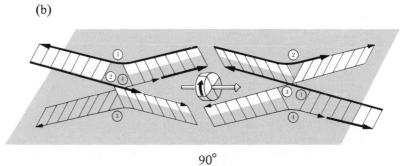

90°
Rotation

(1) Randomly occurring single-strand breaks in one DNA duplex induce breaks at corresponding locations in the opposite duplex. Paired structures then can form in which a strand from each chromatid is hydrogen-bonded to the complementary strand on the homologous chromatid to create two *heteroduplex* regions adjacent to the double exchange of strands [see Figure 3-8(a), panels (i) through (iv)].

(2) Because strands designated with the same sign in the figure are *equivalent,* the crossed-strand exchange structure (iv) can be represented in either of two ways as shown. The two representations are simply two different ways of looking at the same three-dimensional structure, as shown more clearly in Figure 3-8(b).

(3) The double exchange can move along the paired duplex structure by concerted dissociation and reassociation of complementary strands. This process, called *branch migration,* results in formation of longer heteroduplex regions (v).

(4) The four-stranded structure (v) can be resolved again into independent chromatids by a breakage and repair process involving two single-strand cleavages by a nuclease and closure of single-strand breaks by a DNA ligase (WWBH Chapter 17). Depending upon where the single-strand cleavages occur, resolution can yield either Structure (vi) or Structure (vii). The alternative resolutions can be visualized more easily by referring to Figure 3-8(b). Cleavage at Points 2 and 4 yields chromatids that are nonrecombinant for markers outside of the heteroduplex region (vi), whereas cleavage at Points 1 and 3 yields recombinant chromatids (vii).

◀ **Figure 3-8.** (a) A molecular mechanism for recombination between two DNA molecules in nonsister chromatids. of a tetrad. For clarity, the distance between the two duplexes is greatly exaggerated and helical coiling is not shown. The complementary strands of each duplex are arbitrarily designated + and −. (i) A single-strand break occurs in one of the DNA duplexes. (ii) One of the resulting single-strand ends promotes formation of a second single-strand break at the equivalent position in the homologous duplex. (iii) A unitary crossed-strand exchange occurs by displacement of a single strand adjacent to the second break and formation of a heteroduplex region. (iv) The displaced strand forms a second region of heteroduplex, creating a double crossed-strand exchange. This structure can be represented in either of two equivalent ways (see text). (v) Branch migration can move the crossover point in either direction to extend the heteroduplex regions. (vi) and (vii) Resolution occurs by single-strand breakage at the crossover point and repair to give either of two structures (see text). (b) Representation of the three-dimensional structure that results from a double crossed-strand exchange, showing that the alternative forms of (iv) and (v) in (a) simply represent different ways of viewing the structure.

3-8   The behavior of chromosomes in meiosis governs the inheritance of genetic traits

(a)   The basic principles of inheritance were deduced in the 1860's by Gregor Mendel from the results of genetic crosses with pea plants, long before the chromosomal basis of heredity was discovered. Mendel's first law states that the alleles of a given gene segregate at each generation; that is, a heterozygous A/a individual who carries alleles A and a always transmits one of the two, with equal probability, to each offspring. This principle is the direct result of events in the first meiotic division that leads to formation of the individual's gametes. Alleles of the same gene in diploid cells are carried at corresponding loci on two homologous chromosomes. The two homologues segregate during the first meiotic division to separate daughter cells; division of these cells produces two gametes with one homologue and two gametes with the other homologue. Thus each gamete receives one of the two alleles with equal probability.

A cross between two heterozygous A/a individuals, each producing A and a gametes with equal probability, will produce offspring with three different genotypes, AA, Aa, and aa in a 1:2:1 ratio. If A is dominant to a, then the ratio of A to a *phenotypes* in the progeny will be 3:1. The results of such a cross can be predicted conveniently by constructing a grid of the gametic genotypes and filling in the squares as shown in Figure 3-9.

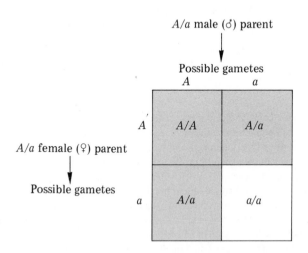

Figure 3-9.   Grid for predicting genotypes and genotype frequencies among progeny of a cross between individuals heterozygous for a single pair of alleles, $A/a \times A/a$. A slash is used to separate the alleles on two homologous chromosomes. Capital and small letters are used to indicate dominant and recessive alleles, respectively. Screened squares on the grid indicate progeny that will show the dominant phenotype.

(b)   Mendel's second law states that unlinked alleles segregate independently. Some pairs of alleles appear to be *linked;* that is, they generally are inherited together. For example, an *AaBb* individual may produce almost entirely *AB* and *ab* gametes, and only rarely *Ab* and *aB* gametes. By contrast, other pairs of alleles appear to be *unlinked;* that is, they are inherited completely independently. For example, an *AaCc* individual may produce *AC, Ac, aC,* and *ac* gametes with equal frequency. We now know that the linkage illustrated in the first case results when the *A* and *B* loci are close together on the same chromosome. In the example given, the linked alleles must be present in the configuration *AB/ab* where the slash separates the markers on the two homologous chromosomes. Unlinked loci are located either on different chromosomes, or far enough apart on the same chromosome that the probability of recombination between them is ~50% (see following section). Since the orientation of the homologous chromosome pairs at the cell equator in meiotic Metaphase I is random, alleles on different chromosome pairs will assort independently among the two daughter cells of the first meiotic division and the resulting gametes. The genotypes and genotype frequencies that result from crosses involving two or more unlinked alleles can be determined by constructing grids from the possible gametic genotypes as in Figure 3-9.

(c)   Occasionally, an individual that is heterozygous for each of two linked alleles, for example *AB/ab,* will produce not only the parental-type *AB* and *ab* gametes, but also recombinant *Ab* and *aB* gametes that result from a recombination event during meiotic prophase (see Figure 3-7). Recombinant gametes can be detected by the appearance of recombinant progeny in an appropriate cross. For example, if *A* and *B* are dominant to *a* and *b,* respectively, then a cross of the *AB/ab* individual to an *ab/ab* individual will produce mostly progeny that show either the dominant (*AB*) or recessive (*ab*) phenotype for both alleles. However, recombinant gametes will produce progeny that can be recognized as showing the dominant phenotype for one allele and the recessive phenotype for the other (*Ab* or *aB*). This type of cross, involving an individual homozygous for all the recessive alleles in the cross, is called a *test cross*. From the results of this test cross, the percent recombination (R) between the *A* and *B* loci can be calculated using the expression

$$R = \frac{\text{recombinant progeny}}{\text{total progeny}} \times 100\% \qquad (3\text{-}1)$$

R for markers on the same chromosome can vary from 0 to 50%. Markers for which R = 50% are defined as unlinked. R cannot exceed 50%, because recombinants for two markers are detected only if an odd number of exchanges has occurred between them. For closely linked markers, the probability of one exchange is far greater than the probability of two or more, but as the distance between markers increases, multiple exchanges become more frequent. For unlinked markers, the probabilities of odd-numbered and even-numbered multiple exchanges are equal, so that R = 50%.

(d) Meiotic recombination is generally reciprocal; that is, a single event produces both possible recombinant chromatids. Therefore, a tetrad that has undergone one recombination event generally consists of two recombinant and two nonrecombinant chromatids (Figure 3-7), each of which is distributed to one of the four gametes produced in the meiosis. For example, in an *AB/ab* individual, a single meiosis involving recombination between the *A* and *B* loci usually produces gametes of the four genotypes *AB, ab, Ab,* and *aB*.

Nonreciprocal recombination events occasionally occur, however, due to a phenomenon called *gene conversion*. This phenomenon has been studied extensively in attempts to understand the molecular mechanism of recombination. Gene conversion operates on genes that happen to be included in the region of DNA heteroduplex that is formed during a crossed-strand exchange [Figure 3-8 (iii)]. Consider the recombinant chromatids arising from such an exchange (Figure 3-10). If one of the allelic markers (*B/b*) analyzed in the cross falls within the region of the original exchange, then the recombinant chromatids will contain regions of heteroduplex in which one strand derives from the *B* and the other from the *b* homologue. Since the genetic difference between *B* and *b* reflects a difference in DNA nucleotide sequence, the duplex must be mispaired at some point in this region. All cells contain repair enzymes that eliminate such mispairings by excision and resynthesis of one DNA strand, randomly chosen, over a short stretch including the mispaired region (WWBH Essential Concept 17-6). In the example shown in Figure 3-10, gene conversion occurs because the *b* allele is retained during repair of both duplexes. The resulting chromatids are *ab* and *Ab*, and the four gametes produced in the meiosis will be *AB, ab, ab,* and *Ab*, indicating an apparently nonreciprocal recombination event.

3-9 Genetic maps are based on the relationship of recombination frequencies to distances between chromosomal loci

(a) To a first approximation, the occurrence of recombination events is random along each chromosome. Therefore the frequency of recombination between two linked allelic pairs is proportional to the physical distance between them. This relationship can be exploited to construct maps of chromosomal loci based on recombination frequencies measured in genetic crosses. By definition, one map unit corresponds to a recombination frequency of 1%. A map of the *Drosophila* genome is shown in Figure 3-11.

(b) Departures from randomness of recombination can occur for two reasons.

    (1) One recombination event can influence the occurrence of another nearby event. Cases of both *positive interference* (suppression of a second exchange by a first exchange) and *negative interference* (promotion of a second exchange by a first exchange) have been observed. Double exchanges between two markers in a genetic cross

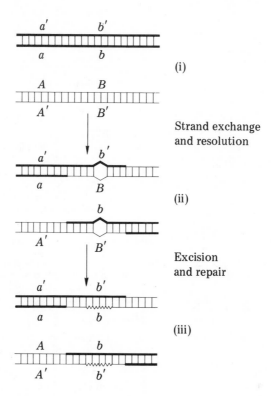

Figure 3-10. A nonreciprocal recombination event resulting from gene conversion. The duplexes in (ii) are derived from those in (i) by the process illustrated in Figure 3-8(a) [(i) → (vii)]. The original reciprocal strand exchange occurs at the *B/b* locus, producing a mismatch in the heteroduplex regions of both chromatids in (ii). Heavy and light lines indicate DNA strands from the two original parental chromatids (i); wavy lines indicate new DNA synthesized in the repair process. Primed and unprimed letters indicate the complementary nucleotide sequences in the two DNA strands corresponding to a particular allele.

result in nonrecombinant chromatids, which must be corrected for in estimating the true frequency of recombination events. Interference introduces uncertainties into such corrections.

(2) Experiments on recombination in microorganisms show that certain regions of DNA duplex recombine more frequently than others, for unknown reasons. Such nonrandomness leads to distortions in the genetic map relative to true physical distances between markers.

In general, however, except over very short intervals, genetic mapping based on recombination frequencies provides a reliable method for estimating relative locations of chromosomal loci.

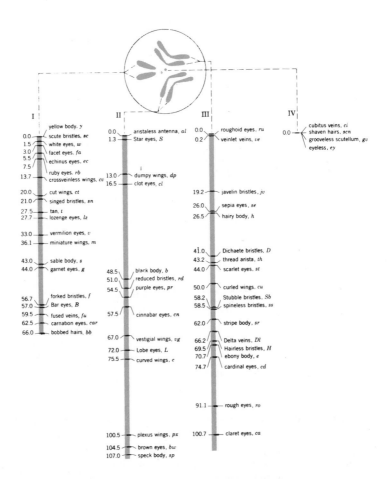

Figure 3-11. Genetic map of the four linkage groups in *Drosophila,* showing their correspondence to the four pairs of homologous chromosomes in a female diploid cell. The units on each linkage group represent map units. (From Eldon J. Gardner, *Principles of Genetics,* Wiley, New York, 1968.)

3-10 Some genes are inherited independently of the nuclear chromosomes

(a) Occasionally, genetic traits are found that show a pattern called *maternal* or *extrachromosomal inheritance,* which is inconsistent with the predictions of Mendelian segregation discussed in Essential Concept 3-8. A simple example is the inheritance of *poky,* a defect in the respiratory apparatus of mitochondria in the bread mold *Neurospora.* When male gametes from a *poky* strain fertilize wild-type female gametes, all the progeny are wild-type, but when male wild-type gametes fertilize *poky* female gametes, all the progeny are *poky.* Since the genetic constitution of the two zygotes should be identical, this result cannot be explained by the action of chromosomal genes. Investigation of this and similar phenomena led to the discovery that mitochondria and chloroplasts contain autonomously replicating genetic determinants that are independent of the nucleus. The inheritance of these determinants in sexual reproduction is determined by the genotype of the organelles in the cytoplasm of the egg cell, and is unaffected by the genetic constitution of the zygote nucleus.

(b) The genes of mitochondria and chloroplasts are carried on small circular DNA molecules (Figure 3-12) analogous in structure and mechanism of replication to the episomes of bacteria (Chapter 1). The mitochondrial DNA in higher animals is about 5μ in length, corresponding to a coding capacity of 15 to 20 gene-equivalents. At least some of the mitochondrial genes code for components of an intramitochondrial protein synthesizing system, including tRNA, rRNA, and possibly ribosomal proteins. This system is distinct from the protein synthesizing system of the cell cytoplasm, and analogous to that of bacteria in both the physical characteristics and inhibitor sensitivity of its ribosomes. It seems likely that the function of this system and possibly of some structural genes in the mitochondrial DNA is to produce some of the highly hydrophobic proteins of the mitochondrial inner membrane, but this notion is not yet proven. The similarity of the genetic systems of mitochondria and chloroplasts to those of bacteria has led to the suggestion that these organelles evolved from procaryotic symbionts in early eucaryotic cells.

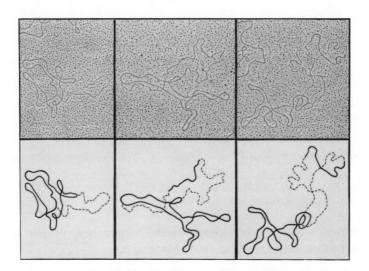

Figure 3-12. Electron micrographs of replicating circular mitochondrial DNA molecules. In the drawings solid lines and dashed lines indicate parental and newly synthesized DNA, respectively. (From D. Robberson *et al., Proc. Natl. Acad. Sci. (U. S.)* **69,** 737, 1972.)

3-11   Additional concepts and techniques are presented in the Problems section

(a)   Phenotypic ratios among progeny in crosses involving independently segregating alleles. Problems 3-7 and 3-9.

(b)   Genetic mapping in *Drosophila*. Problems 3-11, 3-12, 3-13, and 3-18.

(c)   Inheritance of sex-linked characteristics. Problems 3-14, 3-15, 3-17, and 3-18.

(d)   Pedigree analysis. Problems 3-20 and 3-21.

(e)   Tetrad analysis in fungi. Problems 3-24 and 3-25.

(f)   Maternal influences in chromosomal inheritance. Problem 3-26.

REFERENCES

Where to begin

A. M. Srb, R. D. Owen, and R. S. Edgar (Ed.), *Facets of Genetics, Readings from Scientific American,* Part I (Freeman, San Francisco, 1970)

General

J. D. Watson, *Molecular Biology of the Gene* (W. A. Benjamin, Menlo Park, Calif., 1970, 2nd ed.)

Replication of chromosomal DNA

Chromosome Structure and Function: Cold Spring Harbor Symposia on Quantitative Biology, Vol. 38, pp. 195-294 (Cold Spring Harbor Laboratory, Cold Spring Harbor, N. Y., 1973)

Chromosome segregation and the spindle apparatus

R. McIntosh, "Model for mitosis," *Nature* **224,** 659 (1969)

B. Nicklas, "Mitosis," *Adv. Cell Biology* **2,** 225 (1971)

M. Westergaard and D. von Wettstein, "The synaptinemal complex," *Ann. Rev. Genetics* **6,** 71 (1972)

The Biology of Cytoplasmic Microtubules (Symposium N.Y. Acad. of Sciences, in press, 1975)

Chromosomal and extrachromosomal inheritance, genetic analysis, and molecular mechanisms of recombination

S. Emerson, "Linkage and recombination at the chromosome level," in E. W. Caspari and A. W. Ravin (Ed.), *Genetic Organization,* p. 267 (Academic Press, New York, 1969)

H. Kasamatsu and J. Vinograd, "Replication of circular DNA in eukaryotic cells," *Ann. Rev. Biochem.* **43,** 695 (1974)

J. R. Preer, "Extrachromosomal inheritance: hereditary symbionts, mito-chondria, chloroplasts," *Ann. Rev. Genetics* **5,** 361 (1971)

C. M. Radding, "Molecular mechanisms in genetic recombination," *Ann. Rev. Genetics* **7,** 87 (1973)

G. Schatz and T. L. Mason, "The biosynthesis of mitochondrial proteins," *Ann. Rev. Biochem.* **43,** 51 (1974)

N. Sigal and B. Alberts, "Genetic recombination: the nature of a crossed strand exchange between two homologous DNA molecules," *J. Mol. Biol.* **71,** 789 (1972)

A. M. Srb, R. D. Owen, and R. S. Edgar, *General Genetics* Chapters 3 and 6 (Freeman, San Francisco, 1965, 2nd ed.)

H. Stern and Y. Hotta, "Biochemical controls of meiosis," *Ann. Rev. Genetics* **7,** 37 (1973)

H. L. K. Whitehouse, *Towards an Understanding of the Mechanism of Heredity* Chapter 16 (St. Martin's Press, New York, 1969, 2nd ed.)

PROBLEMS

3-1 Answer the following with true or false. If false explain why.

(a) Organisms with different genotypes can have the same phenotype.

(b) Chromosomal DNA synthesis occurs unidirectionally at many simulta-neously replicating growing points.

(c) Meiotic chromosomes may be observed after appropriate staining in nuclei from rapidly dividing epithelial (skin) cells.

(d) During the first meiotic division, the genes on homologous paternal and maternal chromosomes are mixed by genetic recombination.

(e) As each human grows older, repeated genetic recombination events in so-matic cells produce patches of cells with different genotypes.

(f) All of the sperm from one human male are genetically identical.

(g) A single pair of homologous chromosomes may be homozygous for some alleles and heterozygous for others.

(h) An organism with three heterologous chromosome pairs can produce 3^2 or 9 kinds of gametes with different assortments of chromosomes.

(i) The assortment of heterologous chromosomes that a gamete will receive is fixed at the first meiotic division.

(j) Chiasmata are the visible expression of single crossover events.

(k) In the test cross $Ac/aC \times ac/ac,$ in which A and C are dominant alleles, the frequency of different phenotypes in the progeny is exactly the frequency at which gametes are generated in the heterozygous individual. Thus the percent recombinant progeny in such a cross is a direct measure of map distance between the genetic markers.

(l) During meiotic Prophase I, crossover events occur either between paternal and maternal nonsister chromatids or between sister chromatids of the same homologue.

(m)   Among the progeny of crosses between mice, half of the males, on the average, have received their X chromosome from their mother and the other half have received their X chromosome from their father.

(n)   Gene conversion requires the existence of a heteroduplex DNA region in at least one sister chromatid.

(o)   The precise matching of the homologous regions of the DNA molecules during genetic recombination is dictated by the synaptinemal complex.

(p)   Branch migration is an enzyme-catalyzed process.

(q)   A crossed-strand exchange can be resolved in such a way that gene conversion occurs without recombination of outside markers.

(r)   Mitochondria and chloroplasts are autonomous in that all their protein components are coded for and produced by an endogenous genetic system.

3-2   (a)   In *Homo sapiens, Mus musculus* (mouse), and *Drosophila melanogaster,* XX is _____ and XY is _____.

(b)   Synaptinemal complexes are most likely to be observed at the _____ stage of the _____ meiotic division.

(c)   The average number of crossover events between any two homologues in any given meiotic prophase is generally _____.

(d)   The time of occurrence of genetic crossing-over is the _____ stage of the _____ meiotic division.

(e)   _____ and _____ are complementary processes that make up the sexual cycle; asexual cell division occurs by _____.

(f)   Even-numbered genetic exchanges between markers that are far apart reduce the apparent percent recombination to a limiting value of _____.

(g)   A mating of a heterozygous individual with one that is homozygous recessive, to determine the frequency of recombinants, is called a _____.

(h)   If a mutation is expressed when heterozygous but lethal when homozygous, the ratio of phenotypes among the progeny of a mating of two individuals with mutant phenotype will be _____.

(i)   Genetic markers carried on the DNA molecules of mitochondria and chloroplasts show a pattern of transmission called _____ inheritance.

3-3   What stage of meiosis is represented in Figure 3-13, assuming that the cells are the same as those represented in Figure 3-5?

3-4   Draw diagrams of mitosis and meiosis showing the similarities between comparable stages in the two processes. What unique chromosome configurations are observed during meiosis that are not seen during mitosis? (This is a tedious exercise, but very helpful for understanding the mechanics of the two processes.)

3-5   Assuming no recombination, what fraction of a man's sperm will contain the paternal homologue for every chromosome?

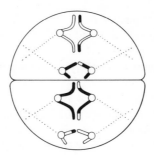

Figure 3-13. A stage of meiosis (Problem 3-3).

3-6 Which of the following treatments might reasonably be expected to stimulate recombination (be recombinogenic)?
(a) UV irradiation before DNA replication
(b) X irradiation
(c) Cross-linking of DNA strands before DNA replication

3-7 (a) Using the grid form shown in Figure 3-9, write the genotypes of the progeny of a cross between parents of genotypes A/A B/B and a/a b/b. (A and B are on separate chromosomes.)
(b) What is the percentage of heterozygotes in the progeny of this cross (the so-called first filial generation, or F1 progeny)?
(c) Using a new grid, write the genotypes of the F2 progeny that will result from a cross between two F1 progeny.
(d) If A and B are dominant to a and b, what is the *phenotypic* ratio in the F2 generation?

3-8 If you know that two sexually reproducing animals, one male and one female, each are heterozygous for one of two unlinked markers (A/a B/B and A/A B/b), how can you determine by genetic crosses between these two animals and their progeny which animal is heterozygous for which marker?

3-9 A *Drosophila melanogaster* with a hairy body (h) was mated to one with a black body (b). The F1 progeny then were mated to produce F2 progeny with the phenotypes shown in Table 3-1.

Table 3-1. Progeny from a *Drosophila* cross (Problem 3-9).

F2 phenotype	Number of progeny
Wild-type	189
Hairy body	72
Black body	58
Hairy black body	19
Total	338

(a)   What were the phenotypes of the F1 progeny?

(b)   Are the genes for hairy and black on the same or different chromosomes, based on the data in Table 3-1?

(c)   Predict the ratio of phenotypes among the progeny from the following matings:

       (i)   hairy parent × F1

      (ii)   hairy parent × F2 hairy black

     (iii)   hairy parent × mixture of all F2 black but not hairy phenotypes

     (iv)   black parent × mixture of all F2 wild-type phenotypes

3-10   Starting with true-breeding (homozygous) *Drosophila* stocks containing one or the other of the two recessive markers, vermilion [*v* — eye color, located on the X (I) chromosome] or rough (*ro* — eye characteristic, located on chromosome III), devise a mating scheme to produce a double homozygous recessive stock.

3-11   In male *Drosophila* a peculiar situation exists: Chromosomes assort independently but no recombination between homologous chromosomes occurs. Thus a male fly that is + +/*ab* can generate only two kinds of gametes, + + and *ab*.

    A female produced in a mating between two flies that each were homozygous for alleles at three loci on chromosome II [dachs (*d*), a leg character; vestigial (*vg*), a wing character; and black (*b*), a body color] was test crossed to a male fly that was homozygous for the three recessive alleles. The data from the cross are shown in Table 3-2.

(a)   What were the genotypes of the F1 female and her parents?

(b)   What is the order of these three loci on the chromosome?

(c)   What is the percent recombination between these loci?

(d)   What would be the result of the test cross if it were F1 ♂ × *d vg b*/*d vg b* ♀?

(e)   What generalization can you make about test crosses in *Drosophila*?

Table 3-2. Progeny from a *Drosophila* test-cross: F1 ♀ × *d vg b*/*d vg b* ♂ (Problem 3-11).

Phenotype			Progeny
+	+	+	9
+	+	*b*	339
+	*vg*	+	71
d	+	+	85
+	*vg*	*b*	79
d	+	*b*	63
d	*vg*	+	373
d	*vg*	*b*	3
		Total	1022

3-12 The F1 progeny from a mating of an ebony (e), rough (ro) female *Drosophila* and a wild-type male are mated. The F2 progeny are listed in Table 3-3.
(a) Can you deduce the percent recombination between the loci for ebony and rough?
(b) What is the special feature of *Drosophila* genetics that allows you to do this?

Table 3-3. Progeny from a *Drosophila* cross (Problem 3-12).

Phenotypes		Progeny
+	+	281
e	ro	86
e	+	19
+	ro	22
	Total	408

3-13 You have just received a stock of *Drosophila* that is homozygous for a recessive allele, virtual (vi), which has been reported to define a fourth autosomal linkage group. As an accomplished geneticist you are convinced that there can be no undiscovered chromosomes in *Drosophila* (there are only three known autosomes), and consequently you are quite skeptical of the report. Armed with three stocks of *Drosophila,* each homozygous for one recessive allele [eyeless (ey), black (b), or hairy (h)] on one of the three autosomes, you set out to prove or disprove the report. Can you devise a set of crosses that will decide the issue unequivocally? (It can be done with six crosses.)

3-14 White (w), eosin (w^e), and wild-type red (w^+) are alleles for eye color in *Drosophila*. A cross between a red-eyed male and an eosin-eyed female gives the progeny shown in Table 3-4.
(a) On which chromosome are these alleles located?
(b) What are the genotypes of the parents?
(c) What is the order of dominance of these alleles?

Table 3-4. F1 progeny phenotypes from a *Drosophila* cross (Problem 3-14).

w^e		w		w^+	
♀	♂	♀	♂	♀	♂
0	279	0	304	565	0

3-15 A small diploid eucaryote recently was found off the coast of California by Thomas Hunts Moregame. It has two sexes (XX, female; XY, male), produces a large number of progeny with a short generation time, will grow on a well-defined medium, and gives rise to mutants with readily distinguishable phenotypes — in short, it is the perfect organism for instructive diploid genetics. This creature resembles a very tiny hippopotamus and consequently has been named *Hippo hypotheticus.*

Professor Moregame has described crosses involving a black male *Hippo hypo,* with the results shown in Table 3-5. Describe as completely as possible the nature of the black allele.

Table 3-5. Progeny of crosses with a black male *Hippo* (Problem 3-15).

1. wild-type ♀ (gray) ✕ black ♂
F1 are all gray

2. F1 ♀ ✕ F1 ♂

Gray		Black	
♀	♂	♀	♂
283	140	0	135

3. F1 ♀ ✕ F2 black ♂

Gray		Black	
♀	♂	♀	♂
158	149	160	153

3-16 You mate four female *Hippo hypos* with wild-type phenotype to males that are homozygous for the recessive alleles earless (*e*), white eye (*w*), and albino (*a*). Table 3-6 lists the phenotypes of the progeny obtained in each cross.
(a) What are the genotypes of each of the female parent strains?
(b) What are the linkage relationships of the three genes?

Table 3-6. Phenotypes of progeny from *Hippo* crosses (Problem 3-16).

Mating	$e\,w\,a$	$e\,w\,a^+$	$e\,w^+\,a$	$e^+\,w\,a$	$e\,w^+\,a^+$	$e^+\,w\,a^+$	$e^+\,w^+\,a$	$e^+\,w^+\,a^+$
1.	0	0	0	203	0	185	210	201
2.	18	173	16	162	162	17	149	19
3.	0	0	0	0	0	0	0	316
4.	0	0	303	0	41	0	29	331

3-17 Wild-type *Hippo hypo* has blue eyes. You have isolated eight independent homozygous mutant stocks that have white eyes (assume that each carries a single mutant allele). The phenotypes of the F1 progeny from all the possible pairwise crosses are given in Table 3-7.

(a) How can you account for the behavior of mutant 5?

(b) Which, if any, of these mutations are sex-linked (located on the X chromosome)?

(c) Excluding mutant 5, how many complementation groups (genes) do your mutants define?

Table 3-7. Phenotypes of progeny from *Hippo* crosses (Problem 3-17). $+$, blue eyes; $-$, white eyes; $+-$, females have blue eyes and males have white eyes; $-+$, females have white eyes and males have blue eyes.

♀ ♂	1	2	3	4	5	6	7	8
1	−	+−	+	+	−	−	−	+−
2	+	−	+	+	−	+	+	+−
3	+	+−	−	−	−	+	+	+−
4	+	+−	−	−	−	+	+	+−
5	−+	−	−+	−+	−	−+	−+	−
6	−	+−	+	+	−	−	−	+−
7	−	+−	+	+	−	−	−	+−
8	+	+−	+	+	−	+	+	−

3-18 A female *Hippo hypo* heterozygous for the recessive alleles yellow eyes (y), curled tail (c), and hairless (h) (the corresponding wild-type alleles are blue, straight, and hairy) is mated to a wild-type male. The progeny from the cross are listed in Table 3-8.

(a) What was the genotype of the original female?

(b) On what chromosome are these alleles carried?

(c) Construct a genetic map for these markers.

(d) Does this represent a general method for mapping loci on this chromosome?

Table 3-8. Progeny from a *Hippo* cross (Problem 3-18).

Phenotype			Progeny ♀	Progeny ♂
+	+	+	1045	6
+	+	h	0	385
+	c	+	0	43
y	+	+	0	79
+	c	h	0	88
y	+	h	0	38
y	c	+	0	410
y	c	h	0	4
		Total	1045	1053

3-19   Humans with three copies of chromosome 13, chromosome 18, or chromosome 21 exhibit the combination of mental and physical defects known as mongolism or Down's syndrome. The frequency of this condition in the general population is about 1 per 1000. Consider a phenotypically normal male whose karyotype shows that one homologue of chromosome 21 has become attached to chromosome 15 in such a way that the centromere of the attached chromosome 21 no longer interacts with the spindle apparatus. The probability of mongolism among this man's offspring will be
(a)   1 in 2
(b)   1 in 3
(c)   1 in 4
(d)   no greater than among the progeny of males with normal karyotypes. Explain your answer.

3-20   This problem and the following one illustrate a method commonly used in human genetics called pedigree analysis.
(a)   Describe the nature of a mutation that is inherited as shown by the human pedigree in Figure 3-14.
(b)   Is there any evidence of aberrant sexual behavior in this family?

3-21   Figure 3-15 is a human pedigree of chondrodystropic dwarfism. Deduce the nature of the mutation causing this condition.

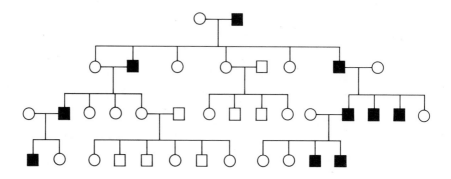

Figure 3-14.   A human pedigree (Problem 3-20). Females are indicated by circles and males by squares. Individuals exhibiting the mutant phenotype are shown by black symbols. Symbols connected by horizontal lines represent mating pairs. Symbols connected by vertical lines to the same horizontal line represent siblings. Generations are represented by vertical lines that connect the horizontal lines.

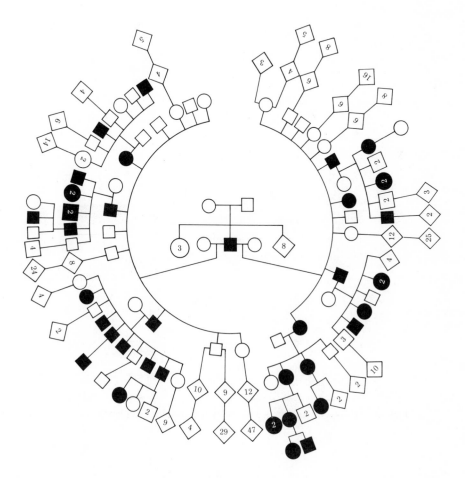

Figure 3-15. A human pedigree showing inheritance of Chondrodystropic dwarfism (Problem 3-21). A female is indicated by a circle, a male by a square. Individuals exhibiting the mutant phenotype are shown by black symbols. A number enclosed by a large symbol indicates siblings who are not listed separately. (Adapted from C. Stern, *Principles of Human Genetics,* Freeman, San Francisco, 1960.)

3-22 (a) Predict whether or not each of the structures in Figure 3-16 will undergo branch migration, and if so, sketch the structure(s) that will result.
(b) What is a general rule governing changes in the number of complementary nucleotide pairs during branch migration?

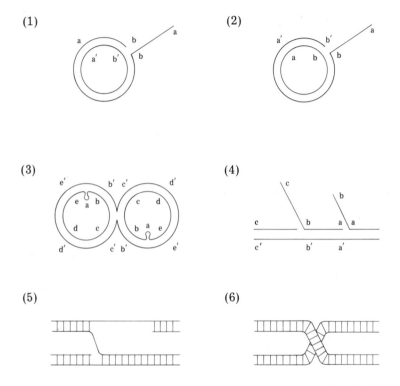

Figure 3-16. DNA structures with and without potential for branch migration (Problem 3-22).

3-23 Draw an equivalent representation for each of the crossed strand-exchange structures in Figure 3-17.

3-24 Much current research in genetics centers around the phenomenon of gene conversion and molecular interpretations of genetic recombination that will account for gene conversion patterns. Although gene conversion apparently operates in all eucaryotes, it is studied most easily in fungi, primarily because the products of individual meioses often are preserved together in *asci* in the form of haploid ascospores as shown in Figure 3-18. The analysis of such ascospores, called *tetrad analysis,* illustrates the power of formal genetics to investigate a detailed molecular mechanism using only observations of gross progeny phenotypes and logical reasoning.

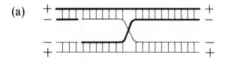

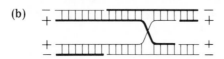

Figure 3-17. Crossed-strand exchange structures (Problem 3-23).

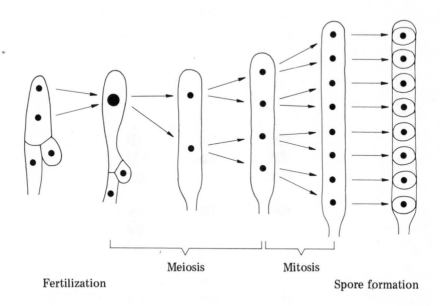

Figure 3-18. Stages of ascospore development in the fungus *Neurospora* (Problem 3-24).

(a) A genetic cross between a normal, black-spored fungal strain and a mutant strain with gray spores (*g*) yielded the types of asci shown in Figure 3-19(a). Which of these asci resulted from simple segregation or from normal recombination and segregation, and which asci resulted from gene conversion? The result of normal recombination on the pattern of ascospore distribution can be predicted from Figure 3-7, if C and c are taken to represent normal and mutant spore color, respectively. Note that the recombinants observed in the fungal experiment indicate exchanges between the color marker and the centromere, rather than between two genetic markers.

(b) Marker genes on either side of the (*g*) site showed about 4% recombination in most matings. However, when aberrant asci with gene conversion patterns were examined, the outside markers showed 36% recombination. Rationalize this observation.

3-25 Leading geneticists of the day were perplexed when, in 1959, asci of the types shown in Figure 3-19(b) were found at low frequency among the progeny of

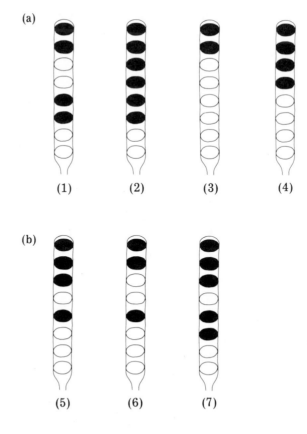

Figure 3-19. Asci with normal and mutant ascospores produced following a fungal cross (Problems 3-24 and 3-25).

crosses such as the one described in Problem 3-24. Can you propose molecular mechanisms that would give rise to these patterns? Keep in mind that a gene-conversion event must begin with a mispaired heteroduplex region, as shown in Figure 3-10, and that ascospore formation involves meiosis followed by mitosis, as shown in Figure 3-18.

3-26   Some phenotypic traits show so-called maternal influences that appear to represent behavior intermediate between normal Mendelian and strict maternal inheritance. An example is the expression of a gene in certain snails that determines whether the coiling of the shell is right-handed or left-handed. The two alleles are called dextral (+) and sinistral (S). Figure 3-20 shows the expression of these alleles through three generations following matings between homozygous right-handed and left-handed parents. The snails are hermaphroditic; that is, they also can reproduce by self-fertilization, so that the F1 progeny can be analyzed without further matings. Formulate a general explanation for the observed patterns of inheritance.

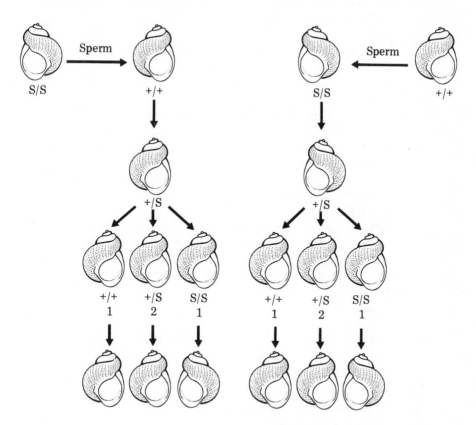

Figure 3-20.   Inheritance of coiling direction in snails (Problem 3-26). (Adapted from A. M. Srb *et al.*, *General Genetics*, 2nd ed., Freeman, San Francisco, 1965.)

ANSWERS

3-1   (a)   True

(b)   False. Replication is bidirectional, and different growing points may be active at different times during the cell cycle.

(c)   False. Mitotic chromosomes may be observed; meiosis occurs only in the germ cells during gametogenesis.

(d)   True

(e)   False. Recombination is thought to be restricted to germ cells. It probably occurs rarely in the somatic cells of higher animals.

(f)   False. Each is likely to be different because each contains not only a random selection of parental chromosomes, but also numerous recombinants between homologous chromosomes.

(g)   True

(h)   False. If no recombination occurs, it can produce 2^3 or 8 different gametes. With recombination, it can generate more, assuming heterozygosity at some loci.

(i)   True

(j)   True (probably; there is still some disagreement on this point).

(k)   True

(l)   False. Crossover events probably occur only between nonsister chromatids.

(m)   False. The X chromosomes in all the male progeny come from the mother.

(n)   True

(o)   False. Synaptinemal complexes direct general pairing; specific pairing is a property of the DNA molecules.

(p)   False. Branch migration is a property of the DNA molecules and requires neither enzymes nor an energy source to occur.

(q)   True. [See Figure 3-8(a).]

(r)   False. The coding capacity of the organellar DNA is not sufficient to code for all of the organellar proteins. A few of these proteins are coded by organellar genes and the remainder by nuclear genes. The latter probably are produced by the cell's cytoplasmic protein-synthesizing machinery.

3-2   (a)   female, male

(b)   pachytene, first

(c)   two or three

(d)   pachytene, first

(e)   Meiosis, fusion (or fertilization), mitosis

(f)   50%

(g)   test cross

(h)   1/3 normal and 2/3 mutant

(i)   maternal (or extrachromosomal)

3-3   The diagram represents late Metaphase II or early Anaphase II.

3-4   The similarities are shown in Figure 3-21. Each stage, except for meiotic Pro-

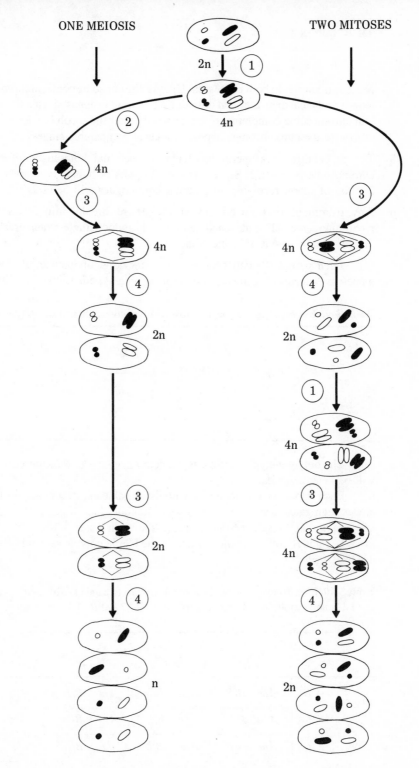

Figure 3-21. A diagrammatic comparison of one meiosis and two mitoses (Answer 3-4).

phase I, is similar in both processes. Unique chromosome configurations such as tetrads and chiasmata are seen only during meiotic Prophase I. In addition, haploid chromosome complements are seen only after the second meiotic division; otherwise the chromosomes appear similar at comparable stages.

3-5 The probability of a sperm receiving the paternal homologue of any one chromosome pair is 1/2. Since there are 23 pairs of human chromosomes, the fraction of sperm receiving all paternal homologues will be $(1/2)^{23}$.

3-6 Any treatment that introduces single-strand breaks into DNA will be recombinogenic. All of the listed treatments introduce single-strand breaks either directly or as a result of replication.

3-7 (a) Each parent (P) can generate only one kind of gamete. P_1 forms AB gametes; P_2 forms ab gametes. The progeny genotype is shown in Table 3-9.

Table 3-9. F1 progeny genotype in a cross involving two unlinked alleles (Answer 3-7).

	P_2 a b
P_1	
A B	A/a B/b

(b) The only genotype possible is the heterozygote A/a B/b; therefore, 100% will be heterozygotes.

(c) Each parent can generate four kinds of gametes, depending on how the alleles segregate during meiotic division. The four kinds of gametes are AB, Ab, aB, and ab. The results of the cross are shown in Table 3-10.

(d) If A and B are dominant to a and b, the F2 phenotypic ratio will be 9 AB:3 Ab:3 aB:1 ab.

Table 3-10. F2 progeny genotypes in a cross involving two unlinked alleles (Answer 3-7). The F2 genotype ratio is 1 $AABB$: 2 $AABb$: 1 $AAbb$: 2 $AaBB$: 4 $AaBb$: 2 $Aabb$: 1 $aaBB$: 2 $aaBb$: 1 $aabb$;

		P_2			
		A B	A b	a B	a b
	A B	A/A B/B	A/A B/b	A/a B/B	A/a B/b
P_1	A b	A/A B/b	A/A b/b	A/a B/b	A/a b/b
	a B	A/a B/B	A/a B/b	a/a B/B	a/a B/b
	a b	A/a B/b	A/a b/b	a/a B/b	a/a b/b

3-8   This problem can be approached best by writing out the results of a cross between the two animals. Half of the F1 progeny will be heterozygous for each marker. The parents can be identified most simply by *back-crossing* each of the F1 progeny to each parent. The parent heterozygous for the *A* marker will yield some *a/a* homozygous recessive progeny; the other parent will yield some *b/b* homozygous recessive progeny.

3-9   (a)   The predominance of wild-type in the F2 progeny indicates that hairy and black are recessive; therefore the F1 progeny were phenotypically wild-type.
(b)   The 9:3:3:1 ratio of phenotypes indicates that the genes for hairy and black are not linked. The two genes must be far apart on the same chromosome or on different chromosomes, but these alternatives cannot be distinguished from the data given. (As indicated in Figure 3-11, *h* is located on chromosome III at 26.5 and *b* on chromosome II at 48.5.)
(c)   (i)   1 hairy:1 wild-type
               (ii)   all hairy
               (iii)   1 hairy:2 wild-type
               (iv)   1 black:2 wild-type

3-10   Mating 1: Cross vermilion ♀ × rough ♂. F1 females are *v*/+ *ro*/+, and F1 males are *v*/Y *ro*/+, in which + indicates the wild-type allele and Y the Y chromosome. Mating 2: Cross F1 ♂ × F1 ♀. Among the progeny will be some vermilion, rough ♂ and ♀ flies that can be bred to produce the doubly recessive stock. Note that mating 2 will not yield ♀ double recessives if mating 1 is rough ♀ × vermilion ♂.

3-11   (a)   In any cross involving linked markers, the most frequent types of progeny will be those produced from parental type (nonrecombinant) gametes. In this example, these gametes must have been + + *b* and *d vg* +; hence the genotype of the F1 female must have been + + *b*/*d vg* +.
(b)   In a cross involving three markers, the least frequent types of progeny will be those produced from double-recombinant gametes that resulted from two crossover events, one between the first and second markers and the other between the second and third markers. Since the two rarest classes are recombinant at the *b* locus, this locus must lie between *d* and *vg*.
(c)   The percent recombination between the individual loci is calculated by summing the progeny that show recombination in the interval and dividing by the total progeny. For the interval between *d* and *b* the recombinant progeny are + + *vg*, *d b* +, + + +, and *d b vg*. The percent recombination for the interval is (146/1022) × 100 = 14.3%. The genetic map for these loci is shown in Figure 3-22.

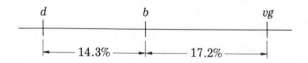

Figure 3-22.   Genetic map of three markers in *Drosophila* (Answer 3-11).

Dachs is not shown in Figure 3-11, but it would lie at about 35 on chromosome II. (d)   If the test cross were an F1 ♂ mated to a homozygous recessive ♀, the progeny would be 50% *d* + *vg* and 50% + *b* +, because there is no recombination in male *Drosophila*.
(e)   Test crosses to determine recombination frequencies in *Drosophila* must be made by mating an F1 ♀ with the homozygous recessive ♂.

3-12 (a)   The recombinants produced in the female are expressed only when they are paired with the male chromosome that carries the recessive alleles. Since this occurs only half the time, the percent recombination is calculated as

$$R = \frac{\text{recombinant progeny}}{\text{total progeny}} \times 100 \times 2$$

These two loci show 20% recombination.
(b)   The absence of crossing-over in the male *Drosophila* allows direct determination of recombination frequencies from phenotypic ratios in the F2 progeny, if the original parents are wild-type and doubly homozygous recessive.

3-13 This question can be answered efficiently by taking advantage of the absence of recombination in male *Drosophila* to test for linkage to each of the three known recessive alleles.
       Cross 1: *vi/vi* × *ey/ey*. If the markers are on different chromosomes, the F1 genotype is *vi/+ ey/+*. If the markers are on the same chromosome, it is *vi +/+ ey*.
       Cross 2: F1 ♀ × F1 ♂. If the markers are on different chromosomes, the F2 will show the characteristic 9:3:3:1 phenotypic ratio, including some *vi ey* progeny. If the markers are on the same chromosome, the F2 ratio will be 2:1:1:0; that is, there will be no *vi ey* progeny.
       Similar crosses using each of the other homozygous recessive stocks will indicate unequivocally whether or not *vi* is located on one of the three known autosomes.

3-14 (a)   Since females in the F1 do not show mutant phenotypes, the alleles must be on the X chromosome (or chromosome I); that is, they are *sex-linked*.
       (b) w^e/w ♀ and w^+ ♂
       (c) $w^+ > w^e > w$

3-15 Black is a sex-linked recessive allele (on the X chromosome).

3-16 (a)   The genotypes of the female parent strains are shown in Table 3-11.
       (b)   Matings (1) and (2) show that the *w* locus is not linked to either the *e* or *a* locus. Matings (2) and (4) indicate that the *e* and *a* loci are linked and recombine with a frequency of 9.9%.

3-17 (a)   The mutant gene in 5 is dominant to the wild-type.
       (b)   The mutations in 2, 5, and 8 are sex-linked.
       (c)   Excluding 5, the mutations define four complementation groups, represented by (1,6,7), (2), (3,4), and (8).

Table 3-11. Genotypes of female *Hippo* strains (Answer 3-16).

1. $e^+\ a/e^+\ a^+$ w/w^+
2. $e\ a^+/e^+\ a$ w/w^+
3. $e^+\ a^+/e^+\ a^+$ w^+/w^+
4. $e\ a/e^+\ a^+$ w^+/w^+

3-18   (a)   The genotype of the original female is $h + + / + yc$.
       (b)   These loci are on the X chromosome.
       (c)   A genetic map can be constructed by considering the data from the male flies only. By the logic explained in Answer 3-11, h must be located between y and c. The map distances are shown in Figure 3-23.
       (d)   This method of mapping generally is applicable to all organisms with sex chromosomes.

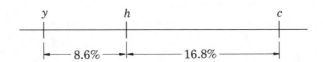

Figure 3-23. Genetic map of three markers in *Hippo* (Answer 3-18).

3-19   When meiosis occurs in this man's germ cells, the fused 15-21 hybrid and the normal homologue of 15 always will segregate to separate daughter cells, either one of which can receive the free homologue of 21 with equal probability. Thus the four kinds of sperm produced will be (1) 15-21, −, (2) 15, 21, (3) 15-21, 21, and (4) 15, −, in equal numbers. Fertilization of a normal egg with sperm of types (1) and (2) will produce phenotypically normal progeny; fertilization with type (3) sperm will produce the mongoloid phenotype; fertilization with type (4) sperm will produce a nonviable zygote containing only one copy of chromosome 21. Hence the probability of mongolism in any child of this man is one in three.

3-20   (a)   The mutation is located on the Y chromosome.
       (b)   Yes. Incest. Four of the offspring on the left side of the pedigree are progeny of a brother-sister mating.

3-21   The mutation is an autosomal dominant.

3-22   (a)   (1)   This structure can undergo migration to give the structures in Part 1 of Figure 3-24.
              (2)   No migration
              (3)   This structure will undergo migration to give the structure in Part 3 of Figure 3-24. This structure is the most stable form because the juxtaposed single-stranded loops relieve stress at the ''joint.''
              (4)   This structure will undergo migration to give the structures in Part 4 of Figure 3-24.
              (5)   This structure will undergo migration to give the structure in Part 5 of Figure 3-24 or intermediate stages. Further migration, which would

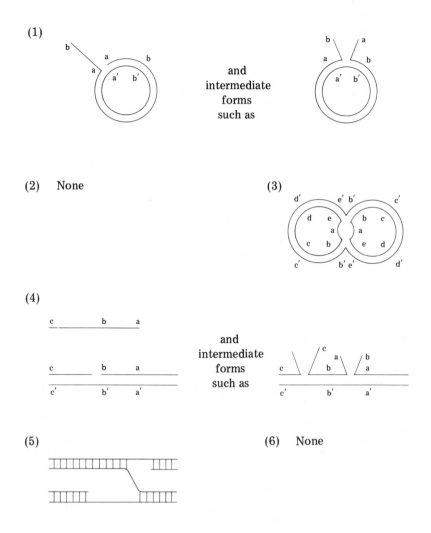

Figure 3-24.   DNA structures resulting from branch migration (Answer 3-22).

result in a net decrease in the amount of duplex structure and a corresponding decrease in stability, will not occur.

(6)   No migration.

(b)   There is no net loss or gain of duplex structure; the dissociation and association of nucleotide pairs are concerted and essentially simultaneous.

3-23   Use the drawing in Figure 3-8(b) to help visualize the equivalent representations. In the absence of such a drawing, the alternative representations can be arrived at by following the general rule that inside strands become outside strands and vice versa: at one end of the structure by rotating each duplex in place, and at the other end by exchanging the positions of two duplexes without rotation. In the answers, shown in Figure 3-25, the duplexes have been rotated at the left end and exchanged at the right end of each structure.

(a)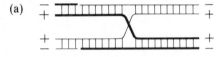

(b)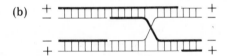

Figure 3-25.   Equivalent representations of two crossed-strand exchange structures (Answer 3-23).

3-24   (a)   Patterns (1) and (4) result from normal segregation, or recombination and segregation. In (4) no recombination has occurred; in (1) a reciprocal recombination has occurred between two chromatids in the tetrad stage. Gene conversion is indicated by apparently nonreciprocal recombination, as seen in asci (2) and (3).

(b)   Since gene conversion occurs in heteroduplex DNA regions generated during crossover events, closely linked markers on either side of the spore color gene are more likely to be recombinant in asci exhibiting gene conversion.

3-25   Consider a strand exchange at the tetrad stage that could lead to gene conversion as illustrated in Figure 3-10. In the mispaired heteroduplex region of each recombinant chromatid, one DNA strand carries wild-type information and the other mutant information, until excision and repair restores complementary pairing by resynthesis in one of the two strands. In ascospore formation, meiotic

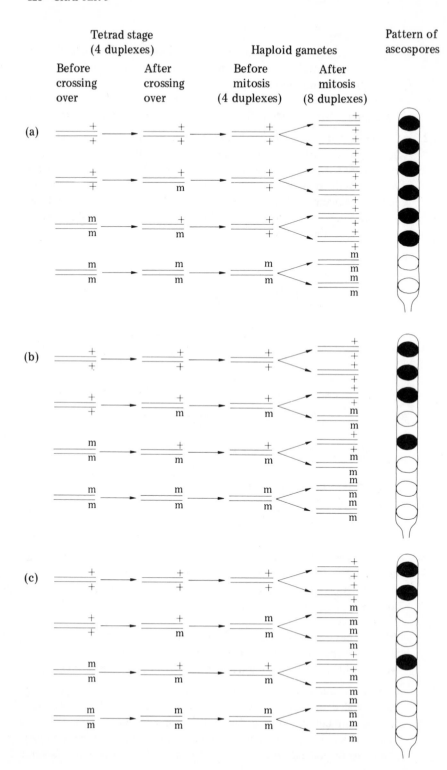

◀ **Figure 3-26.**   The molecular mechanisms of meiotic and post-meiotic segregation of gene conversion events in a fungal cross (Answer 3-25). Horizontal lines represent individual DNA strands; + and m indicate nucleotide sequences corresponding to wild-type and mutant (g) information for spore color, respectively. Reciprocal strand exchanges between two of the four duplexes at the tetrad stage produce mispaired heteroduplex regions containing both + and m information, which may be either repaired (converted) before mitosis to give ordinary gene conversion with meiotic segregation, or replicated without repair during mitosis to give post-meiotic segregation of the + and m markers. (a) Conversion of both heteroduplexes to the + allele before mitosis, giving a 6:2 pattern of ascospores that indicates meiotic segregation (ascus 2, Figure 3-19). (b) Conversion of neither heteroduplex before mitosis, giving an unusual 4:4 pattern that indicates post-meiotic segregation of the two alleles from both recombinant chromatids (ascus 5, Figure 3-19). (c) Conversion of the upper heteroduplex (to m) but not the lower before mitosis, giving a 3:5 pattern that indicates post-meiotic segregation of the alleles from one of the two recombinant chromatids (ascus 6, Figure 3-19). Ascus 7 in Figure 3-19 would be produced in this manner by conversion of the lower heteroduplex (to +) but not the upper before mitosis. From the relative frequencies of meiotic and post-meiotic segregation events in such experiments with the fungus *Sordaria*, it appears that repair before mitosis occurs about 70% of the time.

segregation is followed by a mitosis, in which each chromatid must replicate. If mitotic replication of a mispaired duplex occurs *before* excision and repair, then the resulting daughter duplexes will be wild-type and mutant, respectively. This phenomenon, called *post-meiotic segregation,* is illustrated schematically in Figure 3-26. Pattern (5) in Figure 3-19(b) will result if neither of two reciprocal mispaired heteroduplexes is repaired before replication, and patterns (6) and (7) will result if one of the two mispairings is repaired before replication.

3-26   This pattern of inheritance can be explained by postulating that coiling direction is dictated by determinants in the egg cytoplasm, which in turn depend upon the genotype of the maternal snail. Furthermore, the dextral (+) allele must be dominant to the sinistral (S) allele. Therefore the progeny of individuals with +/+, +/S, and S/S genotypes will be right-, right-, and left-handed, respectively, regardless of either the maternal phenotype or the genotype of the fertilizing sperm.

4 Growth And Genetics Of Somatic Cells In Culture

Genetic studies of eucaryotes such as man are limited by long generation times and small numbers of progeny. In addition, most eucaryotes contain many different cell types whose individual control cannot be analyzed easily in the intact organism. These limitations have been circumvented partially by studying single types of somatic cells growing *in vitro*. The technique of fusing different cell types, with the resultant mixing, segregation, and occasional recombination of chromosomes, has generated a relatively new discipline, somatic cell genetics. This chapter considers the important contributions that somatic cell genetics has made to two areas of eucaryotic biology: mapping of genes and analysis of differentiation.

ESSENTIAL CONCEPTS

4-1 Eucaryotic cells can be grown in vitro

(a)   Adult or embryonic tissues can be dissociated into small fragments by gentle grinding or trypsin digestion. When such fragments are cultured in glass or plastic vessels, fibroblast-like cells migrate out from the fragments and begin to grow on the surfaces of the vessel. These cells continue to divide until they form a continuous sheet of cells one cell thick, which is called a monolayer. Further growth is prevented due to *contact inhibition,* a poorly understood mechanism by which somatic cells in contact inhibit each other's growth. Cell growth can be reinitiated by removing cells from the surface with trypsin and subculturing them at a lower density.

(b)   Mammalian cell growth *in vitro* requires an appropriate temperature, generally 32°C-40°C with an optimum at 37°C; a pH of 7.2-7.4; and added nutrients, including amino acids, vitamins, growth factors, carbohydrates, inorganic salts, lipid, and protein. Table 4-1 lists the minimal defined growth medium for Chinese hamster ovary cells. The metabolic role of the glycoprotein, fetuin, in promoting *in vitro* growth of these cells is unknown. In most cell culture work, serum (e.g., fetal calf serum) is substituted for the lipid and protein components of the defined medium. Mammalian cells can be cultured either as monolayers on glass and certain plastic surfaces or as separated cells in suspension (Figure 4-1).

130

Table 4-1. Composition of F12D, a minimal medium for Chinese hamster ovary cells.

Amino Acids (16)	Vitamins and growth factors (9)	Inorganic salts (9)
Arg	Biotin	KCl
His	Calcium pantothenate	Na_2HPO_4
Ilu	Niacinamide	$FeSO_4$
Leu	Pyridoxine	$MgCl_2$
Lys	Thiamine	$CaCl_2$
Met	Folic acid	$CuSO_4$
Phe	Riboflavin	$ZnSO_4$
Ser	Choline	NaCl
Thr	1, 4-Diaminobutane	$NaHCO_3$
Trp		
Tyr		
Val	**Carbohydrates (2)**	**Lipid (1)**
Glu		
Cys	Glucose	Linoleic acid
Asn	Sodium pyruvate	
Pro		**Protein (1)**
		Fetuin

4-2 Eucaryotic cell growth and division occur in a four-stage cycle

The division cycle of a eucaryotic cell can be divided into four phases, designated G_1, S, G_2, and M (Figure 4-2). *In vitro,* typical mammalian cells divide approximately once every 24 hours.

(1)   A daughter cell emerging from mitosis begins a new cell cycle in G_1 (first gap). As chromosomes are released from their condensed mitotic state, specific regions of the genome become accessible to RNA polymerases, and RNA and protein synthesis resume at a rapid rate. In particular, the enzymes for DNA metabolism and many proteins associated with differentiated functions are synthesized in G_1.

(2)   S is the phase defined by DNA synthesis. The beginning of S is marked by the initiation of replication at many points on the chromosomes. Different chromosomes and chromosomal regions replicate at characteristic times during S (Essential Concept 3-3). In mammalian cells certain chromosomes, such as the Y chromosome in males and the inactivated X chromosome in females, always replicate late.

(3)   During G_2 (second gap) chromosome condensation begins. The physical and chemical bases for condensation are unknown.

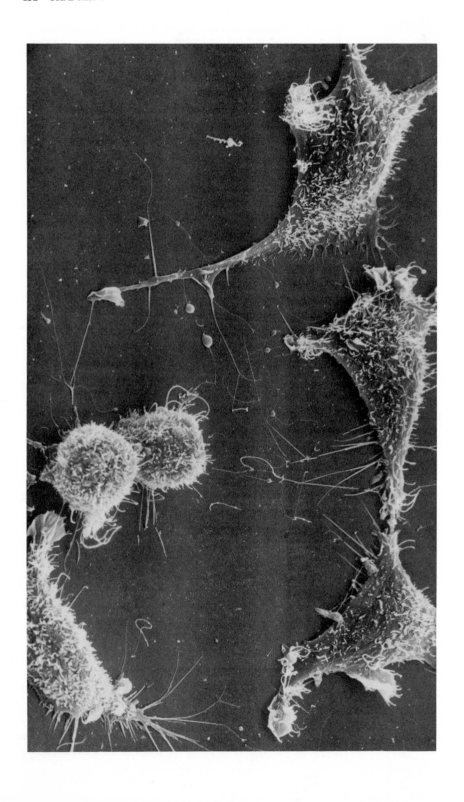

◀ Figure 4-1.   Mouse fibroblasts (L-cells) growing on a petri dish. The two rounded cells have just finished dividing. Photo taken by scanning electron microscopy with 2500x magnification. (Courtesy Dr. J. P. Revel.)

(4)   During M (mitosis) the chromosomes, condensed approximately 10,000-fold, are apportioned equally between the two daughter cells. The nuclear membrane disappears; centrioles migrate to opposite poles of the cell; the spindle apparatus forms; homologues of each chromosome pair are pulled to opposite poles of the cell; the cell membrane pinches off to enclose each daughter cell; and the nuclear membrane and nucleoli reappear (Essential Concepts 3-4 and 3-5).

4-3   Cells in culture can exhibit either a limited or an indefinite life span

(a)   Explants of normal tissues produce cells that display contact inhibition and a normal diploid karyotype. These populations, designated *cell strains,* grow well for a limited number of generations depending on the age and species of the

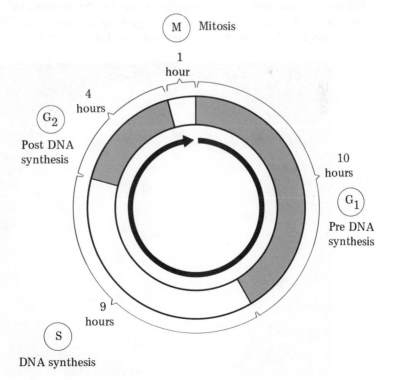

Figure 4-2.   Phases in the life cycle of a typical mammalian cell growing in tissue culture and dividing once every 24 hours.

organism (e.g., 20-30 generations for mouse embryonic cells and around 50 for human embryonic cells). For unknown reasons they then begin to deteriorate and finally cease dividing altogether.

(b)   Cells capable of indefinite proliferation, designated *permanent* or *established cell lines,* can be derived directly from animal tumors. These cells also can arise from cell strains spontaneously or by induction with viral or chemical treatments. Almost all permanent lines are aneuploid. Many have a subtetraploid karyotype and contain chromosomes that have undergone structural rearrangements. Generally, the karyotype of a permanent cell line is variable and is described by the modal (average) chromosome number. With a few exceptions, permanent cell lines, unlike cell strains, no longer display contact inhibition; they grow to very high density and form multiple cell layers. The conversion of a cell strain to a line that is not contact-inhibited is termed *transformation.* Some characteristics of cell strains and transformed and non-transformed permanent cell lines are shown in Table 4-2.

4-4   Most eucaryotic cells can be cloned

If a very few eucaryotic cells are added to a petri dish containing nutrient medium, each will attach to the surface and begin to grow and divide. The progeny from a single cell will form a discrete colony or clone (Figure 4-3). Since all the

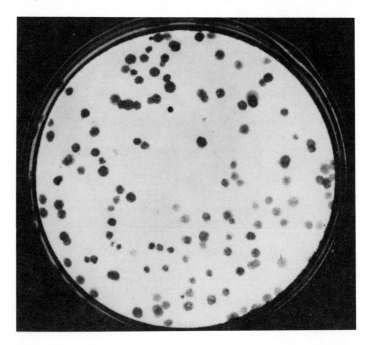

Figure 4-3.   Clones of cultured human cells (HeLa) growing on the agar-covered surface of a petri dish. (Courtesy Dr. T. Puck.)

Table 4-2. Some contrasting properties of cell strains and permanent cell lines. (Adapted from R. Krooth *et al.*, *Ann. Rev. Genetics* **2**, 141, 1968.)

	Cell strains	Permanent cell lines	
		Transformed	Nontransformed
Karyotype	Same as initial karyotype with little or no variation among cells	Different from initial karyotype with considerable variation among cells	Different from initial karyotype with considerable variation among cells
Chromosome morphology	Same as direct squash of donor bone marrow	Often different from direct squash of donor bone marrow	Often different from direct squash of donor bone marrow
Contact inhibition	Present	Absent	Present
Duration of *in vitro* growth	Finite (about 50 generations)	Indefinite	Indefinite
Tissue of origin	Usually benign	Usually malignant	Usually benign
Motility	Motile	Sessile	Sessile
Bioassay for malignancy	Negative	Usually positive in appropriate host	Negative
Frequency with which lines are developed	A strain can be developed from virtually any organ any time	Rare; success unpredictable	Rare; success unpredictable
Cloning efficiency	Generally low	High	High
Growth in suspension	No	Usually	Usually

cells in a colony are derived from one progenitor cell, a single colony can be isolated and used to generate a genetically uniform cell population. This technique is termed *cloning*. Cloning, often in combination with mutagenesis and selection, can be used to isolate homogeneous populations of variant cells from predominantly normal populations.

4-5 Mutant cells can be obtained from genetically defective individuals or by in vitro selection

(a) A number of different mutant cell strains have been isolated by explanting tissue from donors that have a heritable disease. Such cells often carry single-gene defects or chromosomal abnormalities. In addition to their usefulness in genetic studies, such cells permit identification of enzyme defects and their correlation with known diseases.

(b) *In vitro* isolation of eucaryotic cells that carry recessive mutations is complicated by diploidy. A recessive mutant allele that results from mutagenesis will not be expressed as long as the dominant wild-type allele on the homologous chromosome is present. With mammalian cells this difficulty can be circumvented partially by working with genes on the X chromosome, which is effectively haploid in both sexes since one female X is nonfunctional in somatic cells. Haploid strains of amphibian and mouse cells have been produced by artificial *parthenogenesis* (activation of haploid eggs without fertilization), but these strains are quite unstable and paradoxically yield no higher frequency of mutants than ordinary diploid lines. Permanent cell lines, which yield mutants at about the same frequency as cell strains, may compensate for their multiple gene copies (about four on the average) by rapid chromosome loss and duplication. Through chromosome loss a cell may become haploid for a chromosome carrying a recessive mutation, so that the mutant phenotype is expressed. By chromosome doubling in subsequent generations, daughter cells once again may become tetraploid for the chromosome, but each chromosome then will carry the mutation.

(c) In general, mutant cells are isolated from predominantly normal, mutagenized cell populations by selection procedures that favor survival of mutant cells over normal cells. Three frequently used mutagens are 5-bromodeoxyuridine (BrdU), ethyl methane sulfonate, and nitrosoguanidine. One common selection procedure involves treatment of a population of cells with BrdU under conditions that temporarily arrest mutant cell growth but not normal cell growth (Figure 4-4). The normally growing cells incorporate BrdU into their DNA and are killed selectively by subsequent exposure to visible light. Conditions that temporarily arrest only mutant cell growth depend upon the nature of the mutant. For example, high temperature arrests the growth of many temperature sensitive mutants; absence of a nutrient that is nonessential for normal cells selectively arrests the growth of mutant cells that require the nutrient (see Table 4-1).

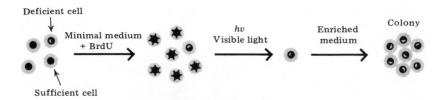

Figure 4-4.   Schematic representation of the BrdU-visible light technique for isolation of nutritionally deficient mutant clones (see text). (From T. Puck, *The Mammalian Cell as a Microorganism,* Holden-Day, San Francisco, 1972.)

4-6   Any two cells in culture can be fused

(a)   Fusion of cells in culture occurs spontaneously, but only very rarely. However, up to 90% cell fusion can be obtained by mixing chemically or UV-killed Sendai virus with concentrated cell suspensions. Sendai and other membrane-coated viruses apparently promote cell fusion by attaching simultaneously to adjacent cells; however, the precise mechanism of their action is unknown.

(b)   Cell fusion yields two kinds of products: cells with multiple but separate nuclei and cells with a single fused nucleus. A cell with two or more nonidentical nuclei is termed a *heterokaryon* (Figure 4-5), whereas a cell with two or more identical nuclei is designated a *homokaryon.* Such cells can remain viable for days to weeks, but unless their nuclei fuse they eventually die without replication. Cells in which nonidentical nuclei have fused to form a single nucleus are termed *hybrids.* In a hybrid cell the genome of each parental cell is functional. In general, hybrid cells can replicate indefinitely.

(c)   Hybrid cells can be distinguished from unfused cells and from cells in which identical nuclei have fused by a variety of techniques.

(1)   The karyotype of a hybrid cell approximates the combined karyotype of the two parental cells (Figure 4-6).

(2)   Unique or *marker* chromosomes from each parent often can be identified in hybrid cells.

(3)   Hybrid multimeric enzymes that contain subunits from both parents serve as hybrid-cell markers if the electrophoretic mobilities of the two parental enzymes differ (Figure 4-7).

(4)   Antigenic markers that are characteristic of each parental cell can be detected on the surfaces of hybrid cells.

(d)   Hybrid cells can be generated from parental cells of the same species (intraspecies cross) or from parental cells of two different species (interspecies cross). Crosses involving two permanent cell lines or a permanent cell line and a cell strain readily produce hybrid cells. In both cases the hybrid cells have the

(a)

(b)

(c)

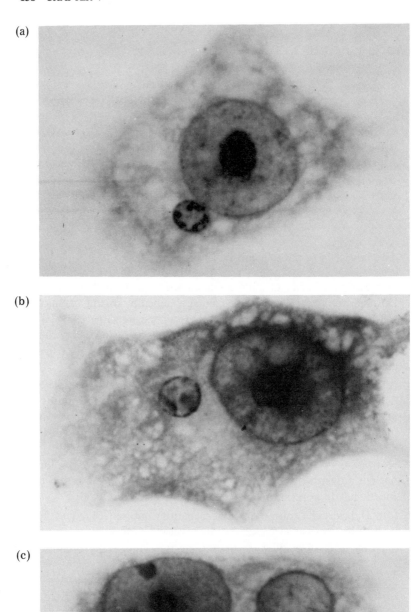

Figure 4-5.   Various stages in the life of a heterokaryon formed from the fusion of a hen erythrocyte (small nucleus) and a human HeLa cell. (a) A recently fused hybrid. (b) A later stage showing enlargement of the erythrocyte nucleus. (c) A still later stage showing still further enlargement. (From H. Harris *et al.*, *J. Cell Sci.* **4,** 499, 1969.)

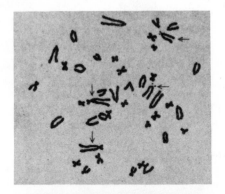

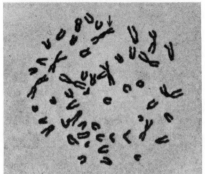

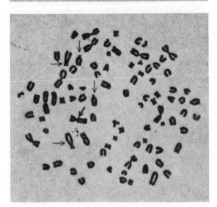

Figure 4-6. Metaphase plates and karyotypes of rat (top), mouse (middle), and hybrid (bottom) cells. Marker chromosomes for the rat are indicated by thin arrows and for the mouse by a thick arrow. (From B. Ephrussi, *Hybridization of Somatic Cells,* Princeton University Press, Princeton, N. J., 1972.)

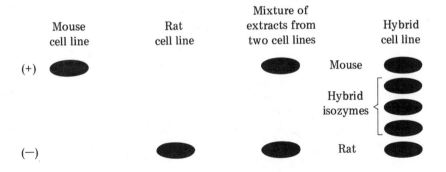

Figure 4-7. Schematic drawing of electropherograms of lactic dehydrogenase patterns from a mouse cell line, a rat cell line, a mixture of extracts from the two cell lines, and a hybrid cell line. In these cell lines the enzyme is a tetramer comprised of identical subunits. The molecules of intermediate electrophoretic mobility are various combinations of mouse and rat subunits. (Adapted from C. Markert and H. Ursprung, *Developmental Genetics,* Prentice-Hall, Englewood Cliffs, N. J., 1971.)

growth characteristics of a permanent cell line. Hybrid cells from crosses of cell strains are very difficult to construct and maintain.

(e)   Most hybrids do not have a selective advantage over the parental cells from which they are derived and therefore are difficult to isolate without special selection procedures. Two such procedures are employed commonly.

(1)   *Full selection* involves fusion of parental cells that are each deficient for a different enzyme, followed by culture in a medium that will support the growth only of cells that produce both enzymes. Under these conditions hybrid cells can grow, but parental cells cannot. One commonly used full-selection procedure involves fusion of parental cells that are deficient in hypoxanthine-guanine phosphoribosyl transferase (HGPRT) and thymidine kinase (TK), respectively. HGPRT catalyzes reactions of the bases hypoxanthine and guanine with 5-phosphoribosyl-1-pyrophosphate to form the corresponding nucleotides IMP (inosine-5'-P, a precursor of AMP and GMP), and GMP, respectively (WWBH Essential Concept 15-3). TK catalyzes the conversion of thymidine to TMP. AMP, GMP, and TMP can be converted to the corresponding deoxynucleoside triphosphates and incorporated into DNA (Figure 4-8). Mutant cells that lack HGPRT can be selected by virtue of their resistance to 8-azaguanine, a guanine analogue that causes cell death when incorporated into DNA. Similarly, cells that lack TK can be selected by virtue of their resistance to BrdU, a thymidine analogue. These en-

zymes are not required by cells under normal culture conditions, since neither enzyme is involved in the major pathway of *de novo* nucleotide biosynthesis (Figure 4-8). However, in a medium containing hypoxanthine, aminopterin, and thymidine (HAT medium), the *de novo* synthesis of GMP and TMP is blocked by the folate antagonist aminopterin. Mutant cells that lack either HGPRT or TK do not survive because they cannot incorporate the nucleosides provided in the medium, but hybrid cells, which contain both enzymes, can bypass the aminopterin block and survive.

(2) *Half selection* involves fusion of one parent cell that carries an enzyme defect with another parent cell that has a very slow growth rate. For example, hybrid cells can be selected from the fusion of a mouse cell deficient in HGPRT with a normal human fibroblast by growth in HAT medium. The parental mouse cell cannot grow in HAT medium and human fibroblasts grow very slowly. However, hybrid cells grow rapidly and form large colonies that are distinguished readily from the slower growing fibroblasts.

4-7 Heterokaryons can be used to analyze nuclear-cytoplasmic interactions

Since heterokaryons can be identified visually, they can be used to study events that take place shortly after fusion. Generally, heterokaryons have been used to analyze the effect that a nucleus or cytoplasm in one state has on a nucleus in another state. For example, in human-chicken heterokaryons a human cell that is

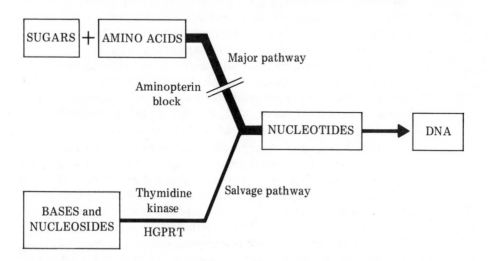

Figure 4-8. The major and salvage pathways for nucleotide synthesis. The major pathway of nucleotide synthesis is blocked by aminopterin. An alternative (salvage) pathway depends on preformed bases and nucleosides and the enzymes TK and HGPRT.

synthesizing DNA and RNA will induce nucleic acid synthesis in the nucleus of a chick red blood cell, which normally does not synthesize DNA and RNA. This experiment suggests that the signals controlling nucleic acid synthesis in these nuclei are quite similar even though the two species diverged more than 200 million years ago. In similar experiments, control of the cell cycle has been studied by analysis of heterokaryons produced by fusion of cells at different stages. For example an S-phase cell can induce premature DNA synthesis in a G_1-phase nucleus and a G_1-phase cell can delay mitosis of a G_2-phase nucleus. Thus there appear to be diffusible factors produced in one cell that can influence cell-cycle control in the other. The molecular nature of these factors is unknown.

4-8   Chromosome losses and translocations in certain hybrids permit gene mapping in somatic cells

(a)   Initially the karyotype of a hybrid is the sum of the parental karyotypes. Intraspecies hybrids maintain relatively constant karyotypes over many generations. In contrast, certain interspecies hybrids always show preferential random loss of chromosomes from one parent. In human-mouse hybrids there is preferential loss of human chromosomes. It is not understood why the chromosomes from just a single parent are lost. It is possible experimentally to influence which parental chromosomes will be lost by introducing breaks into the chromosomes of one parent either with x-ray treatment, or BrdU incorporation followed by exposure to visible light. The fragmented chromosomes selectively are lost.

(b)   Chromosome loss permits genes to be assigned to particular chromosomes. For example, in a human-mouse hybrid that retained only a single human chromosome, any expressed human gene could be assigned to that chromosome. In practice, such one-chromosome hybrids are difficult to obtain, and gene assignments more commonly are made using a series of hybrid clones with limited (1-10) but different combinations of human chromosomes. Generally, analysis of 8-10 such clones allows a correlation to be made between presence of a gene function and a particular chromosome. Using these procedures more than 50 human genes have been assigned to 18 different chromosomes.

(c)   Translocations produce abnormal chromosomes by combining parts of two normal chromosomes. The translocated chromosome or chromosomal fragment can be identified by analysis of fluorescent banding patterns (Essential Concept 2-2). Once a gene has been assigned to a particular chromosome, hybrids with translocations involving that chromosome can be analyzed for the gene product. Correlations between gene function and particular chromosome translocations permit gene assignment to a specific chromosomal region. Familial, spontaneous, and induced chromosomal translocations all have been used in such studies. Familial translocations are heritable abnormalities that are found in certain human families. Spontaneous and induced translocations usually occur subsequent to the formation of a hybrid cell and involve transfer of a portion of a chromosome that is preferentially lost to one that is preferentially retained. Certain viruses induce translocations of particular chromosomes as a result of their tendency to fragment these chromosomes.

4-9 Mammalian cells in culture can express both constitutive functions and differentiated functions

Constitutive or "household" functions, such as those involved in general macromolecular synthesis, the cell cycle, and metabolism, are essential for the maintenance and growth of all cells. *Differentiated* or "luxury" functions, such as the synthesis of hemoglobin, antibodies, and collagen, are necessary for the survival of the organism and the species, but not for survival of the cell.

All cells in culture express constitutive functions. Attempts to establish cultures of normal differentiated mammalian cells have been only partially successful, since such cells often do not express their differentiated functions in culture. However, a number of permanent cell lines derived from tumors do express their differentiated functions. Some examples are melanoma (pigment cell) lines from mouse, hamster, and man; hepatoma (liver cell) lines from rat and mouse; neuroblastoma (nerve cell) lines from mouse; myeloma (antibody-producing cell) lines from mouse and man; and kidney, adrenal, and pituitary lines from the mouse.

4-10 Expression of constitutive and differentiated functions can be studied in cell hybrids

(a)   Hybrid cells derived from parents of different species demonstrate a broad interspecies compatibility of constitutive functions. Membranes fuse with intermingling of proteins and lipids; chromosomal material from different species becomes stably joined by translocation; regulatory signals for macromolecular synthesis from one species turn on nuclei of other species; interspecific hybrids of multimeric enzymes generally are functional. Thus many constitutive functions appear to have been conserved throughout evolution.

(b)   If a cell that expresses one differentiated function is fused to a cell that lacks this function, the resulting hybrid may express the differentiated function fully, partially, or not at all. When the differentiated function is not expressed at all it is said to be *extinguished*. Extinguished differentiated functions occasionally reappear in hybrids after the loss of certain chromosomes from the undifferentiated parent. Thus extinction of differentiated functions probably results from effects of regulatory genes in the undifferentiated parent genome rather than from loss of structural genes from the differentiated parent genome.

(c)   Gene dosage can determine the expression of differentiated functions in hybrid cells. In a permanent cell line the average number of chromosomes, which generally is not a simple multiple of the normal diploid chromosome set, is referred to as the *stemline number,* abbreviated $1s$. If such a cell line undergoes a chromosome doubling, its stemline number becomes $2s$. In certain hybrids of a differentiated and an undifferentiated $1s$ parent cell, the differentiated function is extinguished. However, in similar hybrids made by fusing a $2s$ differentiated cell to a $1s$ undifferentiated cell, the differentiated function is expressed. Moreover, in certain cases the differentiated function is turned on in the undifferentiated cell. Thus gene dosage plays an important role in the regulation of gene expression in hybrid cells.

4-11 Advances in somatic cell culture techniques will raise difficult social and moral questions

(a)   Amniocentesis is a technique for obtaining amniotic fluid from the sac surrounding the early developing human fetus (Figure 4-9). This fluid contains fetal

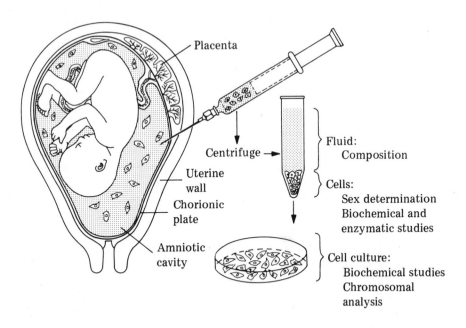

Figure 4-9. Amniocentesis is a technique for prenatal diagnosis that is carried out by inserting a sterile needle into the amniotic cavity and withdrawing a small amount of fluid (see text). An amniotic tap at 16 weeks of gestation provides ample time for diagnostic workup and, if indicated, a therapeutic abortion. (Adapted from T. Friedman, *Scientific American,* November 1971)

cells that can be grown in tissue culture and examined for sex, chromosomal abnormalities, and biochemical defects. This information can be obtained in time for a therapeutic abortion, if indicated. The question of what indications are appropriate for abortion has profound moral, legal, and social implications.

(b)   The improvement of hybridization techniques someday may permit a diploid human somatic cell to be fused with an enucleated egg to form an artificial embryo, which could be reimplanted in a foster mother and borne in the usual manner. Such asexual "cloning" would produce an individual who was genetically identical to the donor of the original diploid cell. Similar experiments have been performed successfully with amphibians. The successful application of this technique to man could affect profoundly our views on human reproduction and the uniqueness of the individual.

4-12   Additional concepts and techniques are presented in the Problems section

(a)   Mechanism of cytochalasin action. Problem 4-3.

(b)   Cell synchronization by a double thymidine block. Problem 4-4.

(c)   Complementation analysis. Problem 4-5.

(d)   Determination of gene linkage using multiply mutant cells. Problem 4-13.

(e)   Mutation rates in somatic cells. Problem 4-18.

REFERENCES

Where to begin

R. Davidson, *Somatic Cell Hybridization: Studies on Genetics and Development,* An Addison-Wesley Module in Biology, Module 3 (Addison-Wesley, Reading, Mass., 1973)

B. Ephrussi and M. Weiss, "Hybrid somatic cells," *Scientific American* (April, 1969)

F. Ruddle and R. Kucherlapati, "Hybrid cells and human genes," *Scientific American* (July, 1974)

General

R. Davidson and F. Cruz (Ed.), *Somatic Cell Hybridization* (Raven Press, New York, 1974)

B. Ephrussi, *Hybridization of Somatic Cells* (Princeton University Press, Princeton, N. J., 1972)

H. Harris, *Cell Fusion* (Harvard University Press, Cambridge, Mass., 1970)

R. Pollack (Ed.), *Readings in Mammalian Cell Culture* (Cold Spring Harbor Laboratory, Cold Spring Harbor, N. Y., 1973)

T. Puck, *The Mammalian Cell as a Microorganism* (Holden-Day, San Francisco, 1972)

J. D. Watson, *Molecular Biology of the Gene* (W. A. Benjamin, Menlo Park, Calif., 1970, 2nd ed.)

Regulation of cell functions

H. Colten and R. Parkman, "Biosynthesis of C4 (fourth component of complement) by hybrids of C4-deficient guinea pig cells and HeLa cells," *Science* **176,** 1029 (1972)

F. Kao, L. Chasin, and T. Puck, "Genetics of somatic mammalian cells X. Complementation analysis of glycine-requiring mutants," *Proc. Natl. Acad. Sci. (U. S.)* **64,** 1284 (1969)

S. Malawista and M. C. Weiss, "Expression of differentiated functions in hepatoma hybrids: High frequency of induction of mouse albumin production in rat hepatoma-mouse lymphoblast hybrids," *Proc. Natl. Acad. Sci. (U.S.)* **71,** 927 (1974)

J. Minna, D. Glazer, and M. Nirenberg, "Genetic dissection of neural properties using somatic cell hybrids," *Nature New Biol.* **235,** 225 (1972)

P. Rao and R. Johnson, "Mammalian cell fusion: Studies on the regulation of DNA synthesis and mitosis," *Nature* **225,** 159 (1970)

Gene mapping

Y. Tan, R. Creagan, and F. Ruddle, "The somatic cell genetics of human interferon: Assignment of human interferon loci to chromosomes 2 and 5," *Proc. Natl. Acad. Sci. (U. S.)* **71,** 2251 (1974)

H. van Someren, A. Westerveld, A. Hagemeijer, J. Mees, P. Merra Khan, and O. B. Zaalberg, "Human antigen and enzyme markers in man-Chinese hamster somatic cell hybrids: Evidence for synteny between the HL-A, PGM$_3$, ME, and IPO-B loci," *Proc. Natl. Acad. Sci. (U. S.)* **71,** 962 (1974)

Social Implications

B. Davis, "Prospects for genetic intervention in man," *Science* **170,** 1279 (1970)

R. G. Edwards and R. E. Fowler, "Human embryos in the laboratory," *Scientific American* (December, 1970)

T. Friedman, "Prenatal diagnosis of genetic disease," *Scientific American* (November, 1971)

PROBLEMS

4-1 Answer the following with true or false. If false explain why.

(a) Because eucaryotes have more genes than procaryotes, eucaryotic mutants are easier to obtain than procaryotic mutants.

(b) Tissue-culture strains usually maintain the differentiated character of the parent cell.

(c) Cell strains can be propagated indefinitely.

(d)   In BrdU selection for mutants in tissue culture, the primary function of BrdU is to increase the mutational frequency.

(e)   Cell fusion is a normal spontaneous process.

(f)   Heterokaryons, like cell strains, have a limited life span.

(g)   Sendai virus promotes cell fusion by actively infecting cells.

(h)   Selection procedures are necessary for isolating hybrids because they generally have little or no selective advantage over the parental cell types.

(i)   Cells that have mutational defects in different genes would be expected to complement one another to produce a more or less normal hybrid cell.

(j)   HeLa cells (a permanent human cell line) have nuclei that synthesize DNA and RNA. Nucleated chicken red blood cells have nuclei that synthesize neither. A heterokaryon of these cell types will show nucleic acid synthesis in the chicken but not in the HeLa cell nuclei.

(k)   Hybrid cells always synthesize the differentiated molecules synthesized by either parent cell.

(l)   All multimeric enzymes that are involved in essential functions must form functional interspecies hybrids; otherwise the hybrid cell could not survive.

(m)   A hybrid constructed from a TK-deficient mouse cell line and a normal human cell line is grown in a medium that requires cells to express thymidine kinase in order to grow. After 100 generations *no* human chromosomes are visible in the hybrid karyotype. The only possible explanation for this observation is that the mouse gene for thymidine kinase has reverted to a functional form.

(n)   Any human genetic disease can be detected by amniocentesis.

(o)   Extinction of a differentiated function in a hybrid cell always results from loss of the chromosome that carries the corresponding structural gene.

(p)   Transformed cells undergo a finite number of cell divisions in tissue culture.

(q)   Hemoglobin, antibodies, and lens crystalline protein are differentiated molecules.

(r)   A typical mammalian cell spends less than 10% of its time in metaphase.

(s)   Cell strains never undergo spontaneous transformation.

(t)   Mammalian cells can be grown in a completely defined medium with no macromolecules.

(u)   Hybrid enzyme molecules must be multimeric.

(v)   Most hybrid enzyme molecules are functional.

4-2   (a)   _____ cell lines divide indefinitely.

(b)   _____ is a process whereby a tissue culture line is made genetically homogeneous (at least initially).

(c)   _____ are single cells containing two or more different nuclei.

(d)   _____ molecules are required by every cell to carry out the cell cycle and macromolecular metabolism.

(e)   If a hybrid cell fails to express a differentiated function from one parent, the differentiated function is said to be _____ .

(f)   Cells that grow to a confluent monolayer and then stop dividing are said to be_____ .

(g)   _____ promotes cell fusion.

(h)   _____ enzyme molecules are multimers composed of one or more subunits from each parent.

4-3   The cytochalasins are a class of drugs that cause cultured cells to extrude their nuclei onto cytoplasmic stalks. This bizarre alteration in morphology is reversible early in the process and apparently is not toxic to the cells. However, with longer exposure to cytochalasin, the nucleus is extruded into the medium leaving behind an enucleated cell. Using the cytochalasins, can you suggest a way to determine whether the mammalian nucleus or cytoplasm is responsible for activating the chicken erythrocyte nucleus in appropriate heterokaryons?

4-4   The cell cycle can be studied by blocking a population of cells at a common point in the cell cycle. When the synchronizing block is lifted, all cells simultaneously begin the cell cycle from the same point. One common technique for synchronizing cells is to block them at mitosis with colchicine (Essential Concept 2-1). Another common technique uses thymidine. An excess of thymidine blocks cells at any point in the S phase. Since the S phase for most cells occupies about 1/3 of the cell cycle, about 1/3 of a random cell population will be blocked at various points throughout S, and 2/3 will be blocked at the beginning of S. In view of this limitation, can you suggest a way to use thymidine to synchronize an entire cell population?

4-5   You have isolated thirteen new Gly-requiring (gly⁻) mutants by BrdU selection. You hybridize each of these new mutants to each of five previously isolated Gly-

Table 4-3. Hybridization analysis of 18 gly⁻ mutants (Problem 4-5). + indicates hybrid growth; − indicates no growth.

Mutant	A	B	C	D	E
1	+	−	+	+	−
2	+	+	−	+	+
3	+	+	+	−	+
4	+	−	+	+	−
5	+	−	+	+	−
6	−	+	+	+	+
7	−	+	+	+	+
8	+	−	+	+	−
9	−	+	+	+	+
10	+	+	−	+	+
11	+	−	+	+	−
12	+	+	+	−	+
13	−	+	+	+	+

requiring mutants, A, B, C, D, and E and test the growth of the resulting hybrids in Gly-free medium. The results are given in Table 4-3.

(a) How many complementation groups are represented among your 18 gly⁻ mutants?

(b) Does your collection of mutants contain any dominant gly⁻ mutations?

4-6 A human diploid cell strain derived from a male with Lesch-Nyhan syndrome (defective HGPRT), when hybridized with a second strain derived from a male with congenital nonspherocytic anemia (defective glucose-6-phosphate dehydrogenase [GPD]), produces tetraploid hybrid cells that express both activities. Since the genes for both HGPRT and GPD are located on the X chromosome and normal human cells contain only one active X, these results are somewhat surprising. Can you offer an explanation for them?

4-7 Human cells lacking enzyme A are fused to mouse cells lacking enzyme B. Hybrid cells are selected in a medium that requires expression of both enzymes. A surviving clone, which grows well in this medium, is found to contain no human chromosomes.

(a) Suggest two explanations for this observation.

(b) How would you distinguish between these alternative explanations?

4-8 Examine the data in Table 4-4.

(a) Propose a "rule" for RNA and DNA synthesis in heterokaryons.

(b) What conclusion about the signals controlling nucleic acid synthesis could you draw from these observations?

Table 4-4. Synthesis of RNA and DNA in heterokaryons (Problem 4-8). 0, no synthesis; 0 0, no synthesis in nuclei of either type; +, synthesis; + +, synthesis in some or all nuclei of both types (From H. Harris *et al.*, *J. Cell. Sci.* **1**, 1, 1966.)

	RNA	DNA
Cell type		
HeLa (permanent human line)	+	+
Rabbit macrophage	+	0
Rat lymphocyte	+	0
Hen erythrocyte	0	0
Cell combination in heterokaryon		
HeLa:HeLa	+ +	+ +
HeLa:rabbit macrophage	+ +	+ +
HeLa:rat lymphocyte	+ +	+ +
HeLa:hen erythrocyte	+ +	+ +
Rabbit macrophage:rabbit macrophage	+ +	0 0
Rabbit macrophage:rat lymphocyte	+ +	0 0
Rabbit macrophage:hen erythrocyte	+ +	0 0

4-9 To assign the genes that code for a series of human enzymes to their chromosomes, you fuse normal human lymphocytes with a permanent mouse cell line lacking TK and select for hybrids by growth in HAT medium. You analyze five hybrid clones for TK and four other human enzymes. Each clone has a different combination of four human chromosomes. The results are given in Table 4-5.
(a) Deduce which chromosome carries the gene for each enzyme.
(b) Which of the enzymes is TK?

Table 4-5. Hypothetical correlation data for human enzymes and chromosomes (Problem 4-9).

		Hybrid clones				
		A	B	C	D	E
Human enzymes	I	+	−	−	+	−
	II	−	+	−	+	−
	III	+	−	−	+	−
	IV	−	−	−	−	−
	V	+	+	+	+	+
Human chromosomes	1	−	+	−	+	−
	5	+	−	−	+	−
	6	−	−	−	+	+
	17	+	+	+	+	+

4-10 The results of an analysis of five human-mouse hybrid clones for six human enzymes are given in Table 4-6.
(a) Deduce which chromosome carries the gene for each enzyme.
(b) These hybrid clones were selected by growth in HAT medium from a fusion of normal human fibroblasts with a permanent mouse cell line that lacked HGPRT. On which chromosome is the gene for HGPRT located?
(c) Your arch rival, Dr. Fuse M. Fast, claims that with these five clones he could have mapped any human gene whose product can be distinguished from mouse products and is expressed in human fibroblasts. Is his boast an idle one? What is the minimum number of clones necessary to uniquely assign any human gene to a chromosome?

4-11 The genes for glucose-6-phosphate dehydrogenase, phosphoglycerate kinase, and HGPRT are located on the X chromosome. From the three translocations involving the X chromosome and their associated enzymes shown in Figure 4-10, define the region of the chromosome in which each gene is located.

Table 4-6. Correlation data for human enzymes and chromosomes (Problem 4-10).

		Hybrid clones				
		A	B	C	D	E
Human enzymes						
1. Phosphopyruvate hydratase		−	+	−	+	−
2. Phosphoglycerate kinase		+	+	+	+	+
3. Thymidine kinase		−	+	+	−	−
4. Nucleoside phosphorylase		−	+	+	+	+
5. Hexoseaminidase A		+	−	+	−	−
6. Hexoseaminidase B		+	−	−	−	−
Human chromosomes	1	−	+	−	+	−
	2	−	+	−	−	−
	3	−	−	−	+	−
	4	−	+	+	−	+
	5	+	−	−	−	−
	6	+	−	−	+	−
	7	+	−	+	−	−
	8	+	+	−	−	−
	9	−	−	−	−	+
	10	+	−	+	+	+
	11	−	+	−	−	+
	12	−	−	+	−	+
	13	−	−	+	−	−
	14	−	+	+	+	+
	15	+	−	+	+	−
	16	+	−	−	−	+
	17	−	+	+	−	−
	18	−	−	−	+	+
	19	−	−	+	+	−
	20	−	+	−	+	+
	21	+	−	−	+	+
	22	+	−	+	−	+
	X	+	+	+	+	+
	Y	−	−	+	+	+

4-12   Eight hypothetical human enzymes, A through H, can be detected by distinct electrophoretic patterns. Mouse L cells lacking TK are fused to normal human lymphocytes and then grown in HAT medium. Nine hybrid clones are grown in HAT medium for 100 generations and then analyzed for the eight enzyme activ-

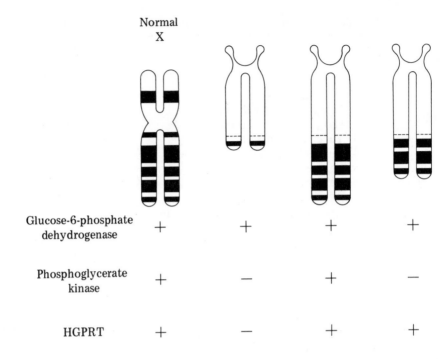

	Normal X			
Glucose-6-phosphate dehydrogenase	+	+	+	+
Phosphoglycerate kinase	+	−	+	−
HGPRT	+	−	+	+

Figure 4-10.   Normal and translocated X chromosomes (Problem 4-11). Shaded areas indicate fluorescent dye banding pattern of human chromosomal material. Dotted lines indicate junction between mouse and human chromosomes.

Table 4-7. Enzyme activity in hypothetical TK$^+$ clones of human-mouse hybrids (Problem 4-12).

Clone	Enzyme activity							
	A	B	C	D	E	F	G	H
1	−	−	+	−	−	+	−	−
2	−	−	+	−	−	−	−	−
3	−	−	+	−	−	+	−	−
4	−	−	+	−	−	−	−	−
5	−	−	+	−	−	+	−	−
6	−	−	+	−	−	+	−	−
7	−	−	+	−	−	+	−	−
8	−	−	+	−	−	+	−	−
9	−	−	+	−	−	+	−	−

Table 4-8. Enzyme activity in hypothetical HGPRT$^+$ clones of human-mouse hybrids (Problem 4-12).

	Enzyme activity							
Clone	A	B	C	D	E	F	G	H
1	+	–	–	–	–	–	–	+
2	+	–	–	–	–	–	–	+
3	+	–	–	–	–	–	–	+
4	+	–	–	–	–	–	–	+
5	+	–	–	–	–	–	–	+
6	+	–	–	–	–	–	–	+
7	+	–	–	–	–	–	–	+
8	+	–	–	–	–	–	–	+
9	+	–	–	–	–	–	–	+
10	+	–	–	–	–	–	–	–

ities (Table 4-7). A second strain of mouse L cells lacking HGPRT is fused to human lymphocytes and grown in HAT medium. Ten hybrid clones are grown in HAT medium for 100 generations and then analyzed for the eight enzyme activities (Table 4-8). Why are certain enzyme activities present in all clones from a given selection, whereas other enzyme activities are present in most, and still others are totally absent?

4-13 A subdiploid line of Chinese hamster ovary cells, designated CHO, has been selected for a number of somatic cell genetic studies because: (1) it grows rapidly with a generation time of 12 hours rather than the 24 hours required by most other cells in culture; (2) the diploid chromosome number is 21 (one chromosome is monosomic) rather than 46 as in human diploid cells; (3) CHO cells with *two* nutritional deficiencies can be isolated by appropriate selective techniques; and (4) hybrids of CHO and human cells readily lose human chromosomes, which can be distinguished from the CHO chromosomes.

As a human geneticist, you see some exciting possibilities for using this system to test the linkage of human genes. Accordingly, you fuse CHO cells that have a double requirement either for inositol and Pro (ino$^-$pro$^-$) or for Gly and Pro (gly$^-$pro$^-$) with normal human cells (ino$^+$pro$^+$). You select hybrid cells in a medium that is supplemented with dialyzed fetal calf serum (in which human cells will not grow) and contains Pro but no inositol or Gly (so that neither CHO cell will grow). After two weeks you test the hybrid clones, which retain five to ten human chromosomes, for the presence or absence of the original Pro requirement (Table 4-9).

(a) How does this experiment test for linkage between the human gene for Pro and the genes for inositol and Gly?

(b) Are the genes for inositol and Gly linked to the gene for Pro?

Table 4-9. Test for linkage between human genes (Problem 4-13). (Adapted from F. Kao and T. Puck, *Nature* **228**, 329, 1970.)

Parental genotypes Human fibroblasts	CHO cells	Pro-containing growth media for hybrids	Number of hybrid clones with in-dicated genotypes	
$ino^+ pro^+$	$ino^- pro^-$	Inositol free	$ino^+ pro^-$	35
			$ino^+ pro^+$	7
$gly^+ pro^+$	$gly^- pro^-$	Gly free	$gly^+ pro^-$	10
			$gly^+ pro^+$	5

4-14   If hen erythrocyte cells are combined with actively dividing cells in heterokaryons, DNA and RNA synthesis are turned on in the hen nucleus (see Problem 4-8). In addition, the hen nucleus enlarges, nucleoli appear, and hen-specific cell-surface antigens begin to be synthesized (see Figure 4-5). The kinetics of the latter events are depicted in Figure 4-11 for heterokaryons between hen erythrocytes and gamma-irradiated mouse fibroblasts (A9 cells) that lack HGPRT. (Gamma-irradiation prevents the active nucleus from undergoing an abortive mitosis that will cause cell death, and thus increases the survival time of the heterokaryons.)

(a)   Describe the sequence of nuclear and cell-surface changes.

(b)   The synthesis of hen HGPRT begins on about the fifth or sixth day fol-lowing hybridization and approximates the time course of hen-specific antigen synthesis. With what nuclear change does antigen synthesis best correlate?

(c)   If a microbeam of UV light is used to destroy specifically the developing hen nucleoli, neither HGPRT nor hen-specific antigen are synthesized. Microbeam destruction of other areas of the nucleus has no effect on hen HGPRT and antigen synthesis. However, nucleoli synthesize only rRNA. Can you offer a suggestion or two as to how nucleoli might affect hen HGPRT and antigen synthesis?

4-15   Selective chromosome loss from one parental genome in a hybrid can be promoted by BrdU incorporation into the DNA of one cell strain followed by ex-posure of the fused cells to visible light. Can you suggest an extension of this technique that might permit selective removal of *specific* chromosomes?

4-16   A graduate student interested in studying genetic control of cell-surface antigens hybridizes cells from a TK deficient mouse line with a strain of normal human diploid cells in HAT medium. She clones a series of individual hybrids and tests them for human surface antigens by determining their relative agglutinability by rabbit antiserum directed against the human cells. A high relative agglutination index indicates that many human cell-surface antigens are present on the hybrid cells, whereas a low index indicates the presence of relatively few human

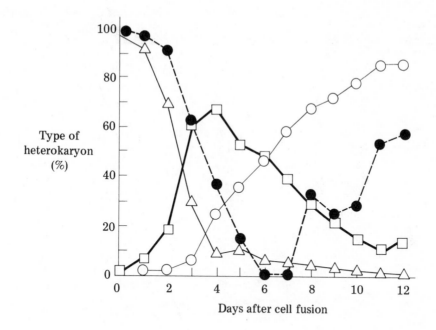

Figure 4-11. Nuclear and cell-surface changes in hen-mouse heterokaryons (Problem 4-14). △ , heterokaryons with unenlarged erythrocyte nuclei; □ , heterokaryons with enlarged erythrocyte nuclei, but not visible nucleoli; ○ , heterokaryons with enlarged erythrocyte nuclei containing visible nucleoli; ● , heterokaryons showing hen-specific surface antigens. (Adapted from H. Harris *et al.*, *J. Cell Sci.* **4**, 499, 1969.)

antigens. The results of these studies are shown in Figure 4-12. Her advisor contends that all the human antigens are coded by a single human chromosome. She thinks he is wrong. Is her advisor's hypothesis sound? Why or why not?

4-17   Tyrosine aminotransferase is produced by a permanent cell line derived from a rat hepatoma (Fu 5-5), whereas it is not produced in another rat cell line derived from liver parenchyma (BRL-1). Tyrosine aminotransferase can be induced by certain steroid hormones in the Fu 5-5 line but not in the BRL-1 line. Hybrid clones derived directly from a fusion of the two lines are not inducible. However, some subclones of original hybrid clones are inducible. The chromosome numbers and baseline and induced levels of tyrosine aminotransferase are given in Table 4-10 for the two parental cell lines, an original hybrid (BF5), and three subclones of BF5.

(a)   One striking observation in these data is that a differentiated function, inducibility by hormone, which was extinguished upon hybridization, can re-

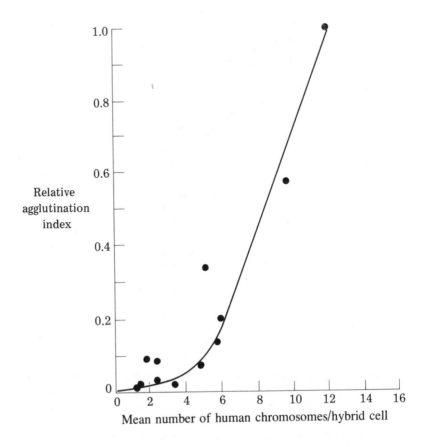

Figure 4-12. Relative agglutination index as a function of the number of human chromosomes in twelve human-mouse hybrid clones (Problem 4-16). (Adapted from M. Weiss and H. Green, *Proc. Natl. Acad. Sci. (U.S.)* **58,** 1104, 1967.)

emerge occasionally upon subcloning. Given the data in Table 4-10, suggest an explanation for this phenomenon.

(b) How many genes appear to be involved in the regulation of tyrosine aminotransferase synthesis?

4-18   Somatic cell genetic studies have used two principal classes of variant cells. One class consists of cells derived from organisms with hereditary defects that are expressed *in vitro* (e.g., galactosemia). The second class consists of variant cells that arise *de novo* in isolated cell populations (e.g., drug resistance, nutritional requirements). Variations that arise *de novo* usually have been interpreted as mutations in structural genes. An alternative explanation is that they result from stable changes in gene expression, similar to processes that occur during embryonic development that do not affect the information content of the DNA.

Table 4-10. Tyrosine aminotransferase in hepatoma hybrids (Problem 4-17). Tyrosine aminotransferase activity is expressed in units/mg protein. (Adapted from M. Weiss and M. Chaplain, *Proc. Natl. Acad. Sci. (U. S.)* **68**, 3026, 1971.)

Cell type	No. of chromosomes	Enzyme activity	
		Baseline	Induced
Parental cells			
Fu5-5	52	22.0	318.0
BRL-1	42	0.27	0.22
Hybrids			
BF5	92	0.78	1.06
BF5-C	92	0.58	0.57
BF5-B	59	0.75	0.8
BF5-1-1	63	0.5	6.0

One approach to distinguishing these alternatives is to test the frequency with which variant cells arise in cell lines at different ploidy levels (numbers of chromosome complements). Table 4-11 lists the frequency with which cells that are resistant to 8-azaguanine arise in diploid, tetraploid, and octaploid male Chinese hamster cells.

(a)   What conclusions can you draw about the rate at which 8-azaguanine resistant cells arise at different levels of ploidy?

(b)   How would you expect the mutation rate for a structural gene to change with increasing ploidy, if the mutation were dominant? If it were recessive and autosomal? If it were recessive and X-linked?

(c)   In view of your answers to Parts (a) and (b), how would you explain the data in Table 4-11? Note that resistance to 8-azaguanine usually is associated with absence of HGPRT activity, a product of an X-linked gene.

Table 4-11. Rate of appearance of 8-azaguanine resistant cells in cells with different ploidy levels (Problem 4-18). Mutation rate is mutations per cell generation. (From M. Harris, *J. Cell. Physiol.* **78**, 177, 1971.)

Cell line	V5 (diploid)	V25 (tetraploid)	V68 (octaploid)
Mutation rate $\times 10^{-5}$	2.2	1.9	2.0
Modal chromosome number	21	43	85

4-19   A clever graduate student decided to study the general subunit structure of a variety of common enzymes by examining hybrid molecules derived from a rat hepatoma-mouse fibroblast hybrid cell line. An electropherogram showing the patterns he observed for several different enzymes is given in Figure 4-13. For each enzyme indicate the general subunit structure and the kinds of subunits that are contributed by the two parental cells (e.g., tetramer, both mouse and rat each contribute two electrophoretically distinguishable subunit forms, A and B) and the subunit composition of each component in the hybrid pattern (e.g., $M^A M^B R_2$, where M and R indicate mouse and rat, A and B superscripts indicate electrophoretically distinguishable subunits, and subscripts indicate number of subunits). Discuss any uncertainties or ambiguities that arise in the interpretation of these patterns.

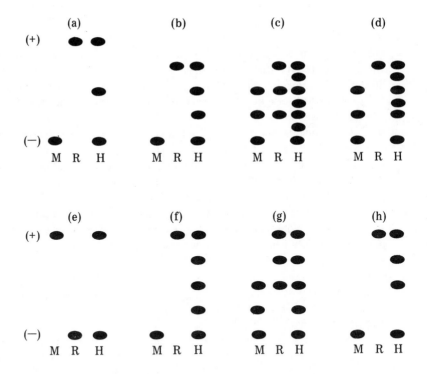

Figure 4-13.   Electropherograms of hypothetical enzymes (a) through (h) in rat-mouse hybrid cells (Problem 4-19). The enzymes are detected by various stains that depend upon activity. M, R, and H indicate, respectively, enzymes derived from mouse, rat, and hybrid cells. No attempt is made to indicate the relative concentrations of the various components of these enzymes.

ANSWERS

4-1 (a)   False. The diploid nature of the eucaryotic genome hides recessive mutations.

(b)   False. Most tissue-culture cells that express a differentiated function are cell lines that are derived from tumors.

(c)   False. Cell strains generally go through 30-50 cell doublings and then die.

(d)   False. BrdU incorporation followed by exposure to visible light is used to kill normal cells.

(e)   True

(f)   True

(g)   False. Sendai virus is chemically or UV killed before use.

(h)   True

(i)   True

(j)   False. Synthesis of RNA and DNA will occur in both nuclei.

(k)   False. Sometimes they do; sometimes they do not.

(l)   False. A certain fraction of multimeric enzymes will be composed entirely of like subunits. For enzymes that form nonfunctional interspecies hybrids (if there are any) the parental-type molecules must be available in quantities sufficient for survival.

(m)   False. A portion of the human chromosome containing the thymidine kinase gene could have been translocated to a mouse chromosome.

(n)   False. The only diseases that can be detected by amniocentesis are those that are characterized by gross chromosomal abnormalities, such as mongolism, or by enzyme abnormalities that are expressed in the amniotic fluid or in the fibroblast, such as Lesch-Nyhan syndrome (lack of HGPRT). One of the hopes for this technique is that appropriate cell fusion will permit human fibroblasts to express functions they normally do not express.

(o)   False. Since differentiated functions sometimes are reexpressed after an initial extinction, the structural gene(s) still must be present.

(p)   False. Transformed cells usually can divide indefinitely.

(q)   True

(r)   True

(s)   False. Spontaneous transformations do occur at a low frequency.

(t)   False. At least one protein is required (Table 4-1).

(u)   True

(v)   True. Most stains that are used to detect hybrid enzyme molecules rely on their biological function. Thus hybrid molecules function *in vitro* and presumably *in vivo*.

4-2 (a)   Permanent

(b)   Cloning

(c)   Heterokaryons

(d) Constitutive

(e) extinguished

(f) contact inhibited

(g) Sendai virus

(h) Hybrid

4-3   You could determine whether the mammalian nucleus or cytoplasm is responsible for activation by first enucleating the mammalian cells, and then fusing them with chicken cells. If enucleated cells activate the chicken cells, the cytoplasm must contain the components necessary for activation. This experiment has been performed. Under these conditions chicken nuclei are activated. This experiment suggests that the mammalian cell cytoplasm is responsible for activation and, furthermore, that the chicken cell's own enzymes for DNA and RNA metabolism are being activated since the mammalian counterparts are not located in the cytoplasm.

4-4   To synchronize an entire cell population, a series of two thymidine blocks are used (Figure 4-14). The first treatment with thymidine brings all cells into the S period. Removal of thymidine for a short time (at least the length of the S period) permits the cells to progress through S. The second thymidine treatment then blocks the entire population at the G_1-S boundary. This process is termed synchronization by a double thymidine block. Excess thymidine blocks DNA synthesis by inhibiting the synthesis of other deoxynucleotides.

4-5   (a) Your collection of mutants contains four complementation groups, I-IV, represented by the following sets of mutants. I: A, 6, 7, 9, 13; II: B, E, 1, 4, 5, 8, 11; III: C, 2, 10; and IV: D, 3, 12.
      (b) All 18 mutations are recessive, since they can be complemented.

4-6   These results indicate that in tetraploid hybrids with two X chromosomes, one from each parental cell, both X chromosomes remain genetically active. This situation contrasts with that in normal female cells, in which one of the two X chromosomes is always inactive. This and similar findings suggest that the ratio of haploid autosomal complements to X chromosomes, rather than the number of X chromosomes *per se,* determines how many X chromosomes remain genetically active.

4-7   (a) The two most likely explanations for this observation are (1) translocation of the human chromosomal region containing the gene for enzyme B to a mouse chromosome and (2) reversion of the mouse B$^-$ gene to B$^+$.
      (b) If the electrophoretic mobilities of the human and mouse B enzymes are distinct, the B enzyme identity in the hybrid cell could be determined. In addition, chromosome banding patterns might demonstrate that an appropriate piece of a human chromosome has been translocated onto a mouse chromosome.

4-8   (a) One rule can account for the data in Table 4-4. If either parent cell normally synthesizes a nucleic acid, synthesis of that nucleic acid will occur in both types

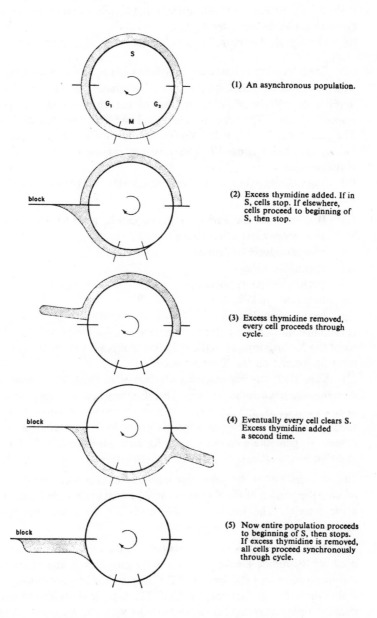

(1) An asynchronous population.

(2) Excess thymidine added. If in S, cells stop. If elsewhere, cells proceed to beginning of S, then stop.

(3) Excess thymidine removed, every cell proceeds through cycle.

(4) Eventually every cell clears S. Excess thymidine added a second time.

(5) Now entire population proceeds to beginning of S, then stops. If excess thymidine is removed, all cells proceed synchronously through cycle.

Figure 4-14.   Double thymidine pulse synchrony (Answer 4-4). [From J. Tooze (Ed.), *The Molecular Biology of Tumor Viruses,* Cold Spring Harbor Laboratory, Cold Spring Harbor, N. Y., 1973.]

of nuclei in the heterokaryon. In no case does the inactive cell suppress synthesis in the active partner. This rule apparently prevails whatever the ratio of the two types of nuclei in the heterokaryon.

(b)   The signals for controlling nucleic acid activation are not species specific.

4-9   (a)   Gene location is analyzed in hybrid cells by correlating gene-product appearance with chromosome presence. The clonal distribution of gene-product appearance will be identical to that of the chromosome carrying the corresponding gene. The genes for enzymes I and III are located on chromosome 5. The gene for enzyme II is located on chromosome 1. The gene for enzyme V is located on chromosome 17. The gene for enzyme IV is not on any of these chromosomes.

(b)   Since all the hybrid clones were selected for the presence of TK, enzyme V must be TK.

4-10   (a)   The genes for the various enzymes can be assigned as follows:

phosphopyruvate hydratase	1
phosphoglycerate kinase	X
thymidine kinase	17
nucleoside phosphorylase	14
hexoseaminidase A	7
hexoseaminidase B	5

(b)   Since these hybrid clones were selected for the presence of HGPRT and since the X chromosome is the only one that is present in all hybrids, HGPRT must be located on the X chromosome.

(c)   Your rival has not made an idle boast. In these five clones each human chromosome has a unique pattern. Therefore any human gene product that can be assayed can be assigned uniquely to a chromosome. In general, the number of unique patterns of plus and minus is given by 2^n, where n is the number of clones. To generate a unique pattern for each chromosome, 2^n must be greater than 24; hence at least five clones are necessary.

4-11   The arrangement of the genes for the three enzymes can be deduced by correlating the portion of the X chromosome present in the translocation with an enzyme activity. The location of the genes for the three enzymes on the X chromosome is shown in Figure 4-15.

4-12   The genes for enzymes C and F probably are on the same chromosome as the gene for thymidine kinase. The genes for enzymes A and H are on the same chromosome as the gene for HGPRT. C may be linked more closely than F to TK since the linkage between F and TK was lost twice (presumably by translocation), whereas C always was found with TK. Similarly, A may be linked more closely than H to HGPRT.

4-13   (a)   Consider the first of the hybridization experiments, in which the gene for inositol was selected for in all hybrids. If the gene for Pro, which was unselected, were present on the same chromosome as the gene for inositol, it would be

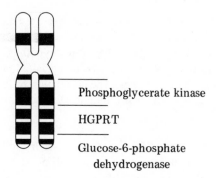

Phosphoglycerate kinase

HGPRT

Glucose-6-phosphate
dehydrogenase

Figure 4-15. Location of genes for glucose-6-phosphate dehydrogenase, phosphoglycerate kinase, and HGPRT on the X chromosome (Answer 4-11).

present in most, if not all, hybrids. If the gene for Pro were on a different chromosome than the gene for inositol, it would be present in a smaller percentage of the hybrids. The expected frequency for an unlinked Pro gene can be estimated from the average number of human chromosomes per hybrid. If one assumes that five different, randomly chosen chromosomes that do not carry the inositol gene are present in the hybrids, one would expect 5/22 or about 23% of the hybrids to contain the gene for Pro.

(b)   Since the gene for Pro is present in only about 17% and 33% of the hybrids that contain the genes for inositol and Gly, respectively, it probably is not linked to either gene.

4-14   (a)   After fusion the hen nucleus first enlarges and then develops nucleoli. Immediately after fusion, hen antigens are present on 100% of the heterokaryons, but during the next six days this value declines to zero. New antigen synthesis begins at the eighth day in erythrocyte nuclei that have nucleoli.

(b)   Hen HGPRT and antigen synthesis correlate best with the appearance of nucleoli.

(c)   Three possibilities as to how nucleoli might affect hen HGPRT and antigen synthesis are: (1) Nucleoli might control mRNA transport out of the nucleus; (2) rRNA might affect mRNA transport out of the nucleus; (3) Hen mRNA may be translatable only on hen ribosomes, which can be made only with functional nucleoli.

4-15   BrdU can be incorporated selectively into, and thereby selectively sensitize, specific chromosomes that replicate at different times in the cell cycle, for example, the late replicating X and Y chromosomes in mammals.

4-16   Her advisor's hypothesis is not sound. Were the human antigens produced from a single chromosome, all the hybrids would have a relative agglutination index of either 1 or 0, depending on whether the hypothesized chromosome was present. The agglutination index should be independent of the number of chromosomes present in the hybrid clones. The correlation of index values with chromosome number suggests that many human chromosomes carry genes for membrane antigens.

4-17   (a)   One subclone, BF5-1-1, reexpresses inducibility by hormone to an extent comparable to the parent Fu5-5 cell. BF5-1-1 has lost a significant number of chromosomes. However, BF5-B has lost even more chromosomes yet retains the capacity to extinguish inducibility. BF5-C, like the original clone, has a full complement of parental chromosomes and the capacity to extinguish inducibility. Taken together these observations suggest that the loss of specific chromosomes permits an extinguished differentiated function to reappear. BF5-B has not lost the chromosome(s) responsible for extinction, whereas BF5-1-1 has. The important conclusion suggested by these observations is that continued extinction of a given tissue-specific function requires the continued presence of some BRL-1 chromosomes.
(b)   There appear to be two separate controls on tyrosine aminotransferase expression: one determines the baseline level and the other inducibility. A change can occur in the expression of inducibility without change in the baseline level of the enzyme.

4-18   (a)   8-azaguanine-resistant cells arise at approximately the same rate at each level of ploidy.
(b)   The mutation rate for dominant mutations should increase with increasing ploidy, since more targets for mutation are available at each increase in ploidy. The rate of appearance of an autosomal or X-linked recessive mutation will be equal to either $(R_M)^n$, in which R_M equals the rate of mutation at an allele and n equals the number of alleles, or more generally $(R_M) (R_L)^{n-1}$, in which R_L equals the rate of loss of a wild-type allele (through mutation, deletion, chromosome loss, etc.). This second expression may be particularly relevant for cell lines that have variable karyotypes. Such cells may express recessive mutations by losing the chromosomes that carry the corresponding wild-type allele (see Essential Concept 4-5). In either case the rate of appearance of a recessive mutation should decline sharply with increasing ploidy.
(c)   Since the observed constant mutation rate does not fit any of the models for structural gene changes discussed in Part (b), it is tempting to speculate that these "mutation" rates actually reflect stable changes in gene expression, which might be expected to remain constant with changes in ploidy. However, stable changes in gene expression cannot account for all 8-azaguanine resistant cells, since some (from diploid lines) have been identified as carrying mutationally altered HGPRT. Two cautionary notes must be added to interpretations of these mutation rates. (1) The structural gene for HGPRT is on an X chromosome. The

results in Table 4-11 are consistent with structural gene alterations in cells that contain only one active X chromosome. However, the bulk of experimental evidence suggests that multiple X chromosomes in multiploids remain active (see Problem 4-6). (2) A simple mutation rate indicates nothing about the various classes of mutation that can lead to an altered cell. In Part (b) it was shown that recessive mutations should decrease and dominant mutations should increase with increasing ploidy. Perhaps the mutations in the diploid line (one X chromosome) are predominantly recessive with only a few dominants, whereas the mutations in the tetraploid (two X chromosomes) and octaploid (four X chromosomes) lines are predominantly dominant mutations with few if any recessives. For example, assume an overall mutation rate of 1.0 for the diploid line with 2/3 of the mutations being recessive and 1/3 being dominant. In the tetraploid and octaploid lines the number of mutants with recessive mutations should be negligible; however, the number of mutants with dominant mutations would double and quadruple, respectively. The overall mutation rates for the diploid, tetraploid, and octaploid lines would be 1.0, 0.7, and 1.3. Thus the overall mutation rate conceivably could be composed of several rates for individual classes of mutations that would sum to an approximately constant overall value, for the three levels of ploidy tested.

4-19   (a)   Dimer; one subunit form each from mouse (M) and rat (R) cells. From $\ominus$ to $\oplus$ the hybrid species are M_2; MR; R_2.

(b)   Trimer; one form each from M and R: M_3; M_2R; MR_2; R_3.

(c)   Dimer; two forms each from M and R: M_2^A; M^AR^A; M^AM^B and R_2^A; M^AR^B; M^BR^A; M_2^B and R^AR^B; M^BR^B; R_2^B.

(d)   Dimer; two forms from M and one form from R: M_2^A; M^AM^B; M^AR; M_2^B; M^BR; R_2.

(e)   Monomer; one form each from M and R: R; M.

(f)   Tetramer; one form each from M and R: M_4; M_3R; M_2R_2; MR_3; R_4.

(g)   Dimer; two forms each from M and R, but M^B is electrophoretically indistinguishable from R^A: M_2^A; M^AM^B and M^AR^A; M_2^B, R_2^A, M^BR^A, and M^AR^B; M^BR^B and R^AR^B; R_2^B. Note that this pattern also is consistent with the absence of functional hybrid enzymes.

(h)   Probably the same as (f), except that the M_3R tetramer is nonfunctional and, accordingly, does not show up on the electropherogram, since the staining depends upon enzyme activity.

5 Eucaryotic Gene Expression

A fundamental problem in the biology of eucaryotes is the understanding of how differential gene expression is programmed to control the development of a single-celled zygote into a multicellular organism composed of many types of cells, each expressing different specialized functions. The problem can be divided arbitrarily but conveniently into two parts. First, what is the mechanism for *determination* of an embryonic cell to follow a particular developmental pathway that will lead to a particular cell type? Second, what is the mechanism for *differentiation* of a precursor cell into a particular cell type in which characteristic specialized functions are expressed? This chapter considers the control of eucaryotic gene expression in determination and differentiation.

ESSENTIAL CONCEPTS

5-1 All the different cell types in an organism arise from one cell, the fertilized egg

(a)   All the genetic information required for constructing an organism is present in the fertilized egg (zygote). With few exceptions, all the differentiated somatic cells of an organism probably contain the same genetic information. This conclusion is based in part on the finding that nuclei of differentiated somatic cells in amphibia can give rise to normal animals when transplanted into enucleated eggs. Therefore the differentiation of a precursor cell into different types of progeny cells is accomplished by selective expression of different subsets of the information in the genome. In this manner, a single zygote gives rise to hundreds of different cell types during an organism's development.

(b)   Formation of specific cell types during development occurs in two major stages: cell determination and cell differentiation. Determination occurs during embryogenesis when a cell becomes committed to eventually expressing a particular phenotype. At each subsequent division, this determined state is transmitted to the daughter cells. Determined cells may divide for many generations before receiving a signal to differentiate, that is, to express the predetermined phenotype. Both determination and differentiation presumably involve specific regulation of gene expression.

5-2   During development, cells become committed to more and more specific developmental subprograms

Development of a eucaryotic embryo involves an irreversible sequence of gene expression that normally continues to completion once it is set in motion. The zygote and early embryonic cells have the potential to become committed to any developmental program. A particular program is initiated by the commitment of a cell to a broad path of development, such as formation of either ectodermal, mesodermal, or endodermal progenitor cells. Subsequent progeny cells become committed to more specific subprograms for functions characteristic of individual tissues or organs (Figure 5-1). In general, normal cell division always produces daughter cells that have a level of determination identical to or more specific than the parental cell.

As an example, the developmental pathway for hemoglobin-producing red blood cells (erythropoiesis) is illustrated in Figure 5-2. The zygote has the potential to generate all cell types. After seven to eight cell divisions, one division gives rise to a precursor (hematopoietic) stem cell that is committed to produce the three major types of blood cells: white blood cells (granulocytes), platelet-producing cells (megakaryocytes), and red blood cells (erythrocytes). One progeny cell of the precursor stem cell subsequently divides to produce an erythroid stem cell, which can proliferate and differentiate into a mature hemoglobin-producing erythrocyte.

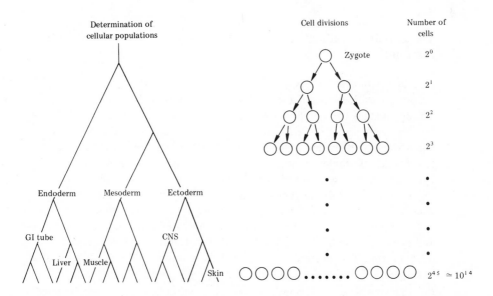

Figure 5-1.   Differentiation and cell proliferation in development. An average of 45 successive cell divisions are required to produce the 10^{14} cells in an adult human.

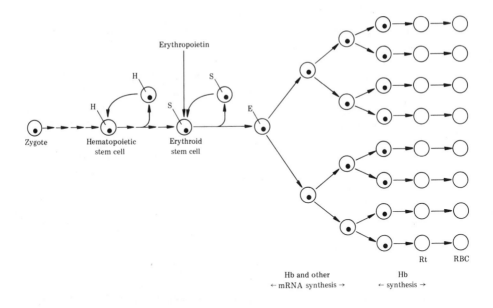

Figure 5-2.   Stages in the development of red blood cells (erythropoiesis). The hormone erythropoietin stimulates differentiation of an erythroid stem cell into an erythroid cell, which begins to synthesize mRNA for hemoglobin (Hb) and, after about four successive cell divisions, gives rise to reticulocytes (Rt, immature red blood cells), which develop into erythrocytes (RBC, mature red blood cells).

5-3   The initial determination of a cell may depend upon its position in the embryo

(a)   In the first few cleavages of a single-celled embryo, the egg cytoplasm becomes partitioned among the cells (blastomeres) of the developing blastula. The egg and the early embryo exhibit two poles that are distinguishable by their morphology and pigmentation. Early cleavages usually occur with a characteristic orientation to these poles. In many organisms the developmental fate of each blastomere is at least partially determined by the portion of the egg cytoplasm that it receives. In several species surgical transplantation of cytoplasmic matter from one region of the egg or early embryo to another region has been shown to change the developmental fate of blastomeres that arise in the altered regions. These observations imply that cytoplasmic components in the embryo are responsible for the initial determination of cells in early development. The molecular nature of these determinants and the mechanisms that specify their positions in the cytoplasm are not known.

(b)   The positions of primordial stem cells in the early embryo were mapped by classical embryologists, who used a variety of techniques to trace cell lineages during development. These studies established that precursor cells for various adult systems occupy specific locations in the embryo, and that patches of several cells can become determined simultaneously to follow a particular developmental pathway. These conclusions can be illustrated by experiments with two recently developed systems, *Drosophila* mosaics and allophenic mice, which promise to be important in future work on development.

(1) An elegant technique for mapping the positions of various primordial stem cells in the early *Drosophila* embryo has been developed using *gynandromorphs,* that is, mosaic flies that consist of both female (X/ X) and male (X/O) cells. Such flies are found among the offspring of parents that carry an abnormal form of the X chromosome, called ring-X (X°). In X/X° female zygotes, for unknown reasons, a ring-X chromosome often is lost during an early nuclear division, so that one of the resulting daughter nuclei becomes X/O and the other remains X/X° [Figure 5-3(a)]. In subsequent divisions, the ring-X is retained in the cells descended from the X/X° nucleus, so that both X/O and X/X° cells are present in the adult fly.

Formation of the blastula in *Drosophila* proceeds by a series of nuclear divisions, followed by migration of nuclei to the surface of the embryo [Figure 5-3(b) and (c)]. The orientation of the spindle in the first nuclear division is random. During subsequent divisions and migration, the nuclei retain their relative positions. For gynandromorphs, the result is that the plane of demarcation between X/X° and X/O cells also is randomly oriented in larvae and adult flies [Figure 5-3(d)-(f)]. If the normal X chromosome carries a recessive gene that alters body color, the outline of this plane can be seen from the difference in pigmentation of the X/X° and X/O cells. The probability that the demarcation line will fall between two sur-

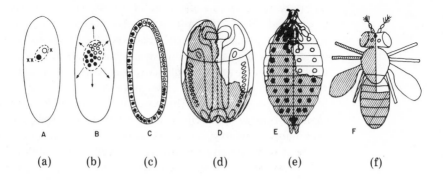

(a)	(b)	(c)	(d)	(e)	(f)

Figure 5-3. Formation of a gynandromorph mosaic fly, showing (a) one X chromosome lost during the first nuclear division after fertilization; (b) nuclei migrating to the surface of the embryo; (c) composite blastula of X/X° and X/O cells; (d) fate map of the embryo; (e) composite larva; (f) mosaic fly after metamorphosis. (Courtesy Dr. S. Benzer.)

face features of the fly is proportional to the distance between the embryonic cells in the blastula that give rise to these features. By scoring large numbers of gynandromorphs, these probabilities can be determined and used to construct a two-dimensional *fate map* of primordial stem cells in the embryo, just as recombination probabilities between genetic markers are used to construct one-dimensional genetic maps. Fate mapping shows that the precursor cells for the various structures of the adult fly occupy precisely determined positions *(foci)* in the embryo.

(2)   Gynandromorph analysis is also a powerful tool for determining which tissues are responsible for physiological functions that are altered by mutations on the X chromosome in certain developmental and behavioral mutants. For example, a mutation called *arhythmic,* which eliminates the normal circadian activity pattern in males but not in females, was shown to affect the head of the animal but no other body parts, by analyzing large numbers of gynandromorphs that carried *arhythmic* and a body color marker on the normal X chromosome. With only a very few exceptions, there was a perfect correlation between expression of the *arhythmic* phenotype and male tissue in the head region. The few exceptions probably indicate that the mutational lesion is in the brain. Since the brain is derived from embryonic precursor cells that are close to but distinct from the precursor cells for the head surface, the two tissues occasionally exhibit different genotypes in a gynandromorph.

(3)   Studies with another elegant experimental system, the allophenic mouse, have shown that several cells in the embryo can be determined independently and identically. Allophenic mice are produced by isolating and combining two genetically different embryos at the 8-cell stage, before blastomeres have become determined [Figure 5-4(a)]. When the blastula membranes are removed by proteolytic enzyme treatment, the blastomeres of two embryos will aggregate and mix spontaneously. With a success rate of about 50%, the aggregate embryo, upon reimplantation in a properly prepared female mouse, undergoes a reduction in size followed by normal development. The resulting mouse, called a genetic *chimera,* is made up of cells with two different genotypes, one from each embryo. Phenotypic markers contributed by each set of parents can be used to identify the two types of cells in the adult.

Most tissues contain both cell types, indicating that the tissue must have arisen from two or more progenitor cells that were committed independently to similar or identical developmental programs. For example, allophenic mice derived from a black and white set of parents can show up to 34 alternating black and white stripes of coat color [Figure 5-4(b)]. This observation suggests that 34 cells

(a)

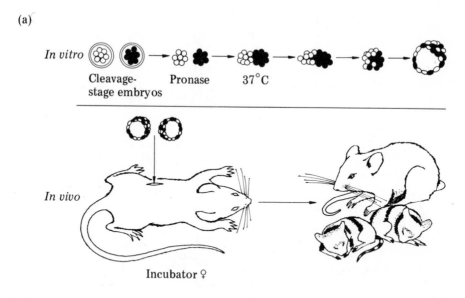

(b)

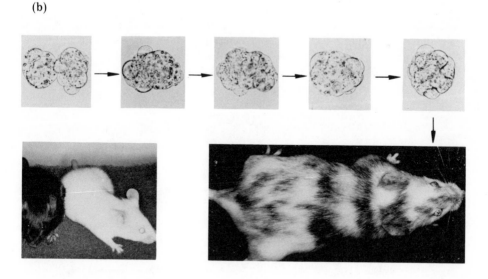

Figure 5-4. (a) Diagram of procedure for producing allophenic mice from aggregated embryos. (b) Parental mice, fusion of embryos, and an allophenic mouse exhibiting segregation of pigment-forming cells (melanoblasts) during development. [From B. Mintz, "Allophenic mice of multiembryo origin," in J. C. Daniel, Jr. (Ed.), *Methods in Mammalian Embryology,* Freeman, San Francisco, 1971.]

along the neural crest of the embryo may be independently determined to be pigment cell precursors (melanoblasts), and that their descendents maintain defined topological relationships to each other during development. The 34 stripes result when the original 34 cells are of alternating genotypes.

5-4   The determined state is generally heritable and stable

(a)   Stability of the determined state is demonstrated clearly in *Drosophila* development. Most structures of the adult fly are formed from discs of determined cells called *imaginal discs,* which are present in the larva but have no apparent larval function. Each disc occupies a specific position in the larva and gives rise to a specific adult structure during metamorphosis (Figure 5-5). Moreover, individual cells within a disc are committed to form characteristic parts of the adult structure.

Although imaginal disc cells in the larva show no signs of differentiation, they are committed rigidly to specific developmental pathways. For example, a wing disc can be removed from one larva and transplanted into almost any location in a second larva. The first larva will metamorphose into an adult lacking one wing, whereas the second larva will metamorphose into an adult with a third wing.

(b)   The determined state of an imaginal disc cell is inherited by all its descendents. From the time of disc determination to the final larval stage, each imaginal disc multiplies from about 10 to 40 cells to several thousand cells. Despite this proliferation, each disc maintains its determined state and differentiates appropriately during metamorphosis.

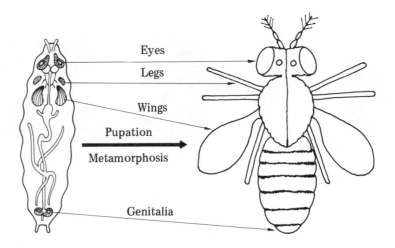

Figure 5-5.   Development of *Drosophila* adult body parts from embryonic imaginal discs.

The stability of the determined state of imaginal discs is illustrated even more dramatically by discs that have been cultured for years by serial transplantation in the abdomens of adult flies, where they proliferate but do not differentiate. Upon implantation into a larva, fragments of such discs still metamorphose into the characteristic adult tissue.

5-5   Occasional changes in determined state suggest relationships between developmental programs

(a)   Occasionally after long cultivation in adult flies, an imaginal disc that is committed to one developmental program will produce a tissue that is characteristic of a different program. This phenomenon is called *transdetermination*. It suggests that one entire program can be replaced by another as the result of a relatively simple switching event.

(1)   Transdetermination occurs in the disc region that is dividing most rapidly, and appears to take place in groups of contiguous cells rather than in single cells.

(2)   Transdetermination is an all or nothing phenomenon. If a cell is transdetermined, it will express only those features that are characteristic of the new determined state. Cells never are transdetermined partially. The new determined state is heritable and as stable as the initial state.

(3)   Transdeterminations from one state to another are asymmetric; that is, the forward and reverse transdeterminations have characteristic and unequal frequencies. Thus when sequential transdeterminations occur, the states tend to follow reproducible pathways (Figure 5-6).

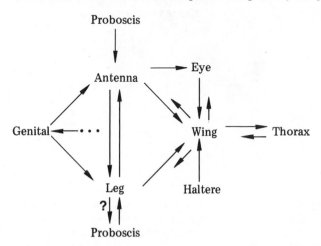

Figure 5-6.   Pathways of transdetermination among the imaginal discs of *Drosophila melanogaster*. Lengths of arrows represent relative frequencies of transdetermination. (Adapted from S. A. Kauffmann, *Science* **181,** 310, 1973.)

(b)   A phenomenon analogous to transdetermination is observed in homoeotic mutants of *Drosophila,* in which there is an abnormal substitution of one adult body part by another. For example, *antennapedia* mutants develop a leg on the head in place of one antenna (Figure 5-7). Most homoeotic mutations are located in a small region on the third chromosome. Analysis of homoeotic mutants has shown that single-gene defects can switch one developmental program to a second, again suggesting that the developmental switches may be simple.

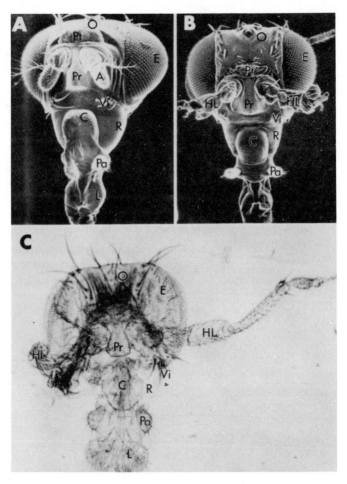

Figure 5-7. Heads of normal *Drosophila* and antennapedia (Antp) mutants. (a) Scanning electron micrograph of a normal head with normal antennae; (b) Scanning electronic micrograph of an Antp head; (c) Photomicrograph of the head of an Antp individual with a large homoeotic leg replacing the antenna. Letters on the photographs indicate antenna (A), clypeus (C), eye (E), labellum (L), ocelli (O), maxillary palpus (Pa), prefrons (Pr), homoeotic leg (HL), ptilinum (Pt), Rostralhaut (R), and vibrissae (Vi). (From J. H. Postlethwait and H. A. Schneiderman, *Developmental Biology* **25,** 606, 1971.)

5-6   Gene expression can be regulated at many different levels

Understanding determination and differentiation at the molecular level requires knowledge of the mechanisms by which gene expression is regulated in eucaryotes. Gene expression can be regulated at several different levels, as indicated in Table 5-1. In general, expression of a gene results in the synthesis of a specific functional protein. For this process to occur, the following requirements must be met: (1) The gene must be activated so that transcription occurs. (2) The mRNA transcript must be transported to the cytoplasm. This step may involve *processing* of the mRNA, that is, modification of its covalent structure. (3) The mRNA must be translated on cytoplasmic ribosomes. (4) The resulting polypeptide must assume its unique three-dimensional structure and, in many cases, undergo covalent modification to a functional form. Some of the mechanisms of gene control given in Table 5-1 are universal, whereas others may occur only in certain cells or organisms.

5-7   In a few systems gene regulation involves changes in genetic information content

(a)   During oogenesis, egg cells are provided with large numbers of ribosomes in preparation for protein synthesis following fertilization. Some organisms meet the demand for rapid rRNA synthesis during oogenesis by amplification of the ribosomal genes, that is, selective replication of the DNA sequences that code for rRNA (Essential Concept 2-15). For example, in *Xenopus laevis* oogenesis, circular DNA molecules consisting of one to fifteen rRNA genes are produced in the nucleolus, a spherical nuclear body that corresponds to the chromosomal region containing the rRNA genes. These DNA molecules replicate, probably by a rolling-circle mechanism (WWBH Figure 17-5), to produce a thousand or more extrachromosomal nucleoli, which migrate to the periphery of the nucleus. The result is an increase of about 1000 fold in the normal complement of rRNA genes at the time of maximum rRNA synthesis.

  Although gene amplification appears to be a convenient way to increase the rate of synthesis of a specific RNA transcript, this mechanism apparently is used only for regulation of rRNA synthesis. Selective amplification of structural genes that code for proteins has never been observed.

(b)   Some cells may become committed to synthesis of specific proteins through translocation and fusion of previously distinct genes. This mechanism has been proposed to account for the differentiation of antibody-producing cells in the immune system. Each polypeptide in an antibody molecule appears to be the product of two distinct genes, one that codes for the C-terminal portion (constant region), which is the same for many antibody molecules, and one that codes for the N-terminal portion (variable region), which is unique to the antibody produced by a given clone of cells. These genes are thought to be joined,

Table 5-1. Possible points of control for gene expression.

Level of control	Possible controllable event	Example
Chromosome	Specific chromosomes or chromosomal regions are inactivated by condensation to heterochromatin during development.	The inactive X chromosome in female mammalian cells is heterochromatic.
	Specific chromosomal regions may be activated by translocation.	Genes for the variable and constant regions of antibody molecules may be joined by translocation before they are activated.
	Specific chromosomes or chromosomal material may be lost during development.	The somatic cells of *Ascaris* lack certain chromosomes.
Gene (transcriptional controls)	The presence of many identical or nearly identical genes in the haploid complement (gene reiteration) permits the corresponding transcripts to be made at a high rate.	*Xenopus* has about 900 ribosomal genes in each somatic cell.
	The extrachromosomal replication of ribosomal genes (gene amplification) permits more rapid production of rRNA.	The *Xenopus* ribosomal gene complement is amplified 1000-fold in the oocyte.
	Specific genes can be activated for transcription.	Different tissues contain different populations of mRNA's.
	Gene activation is accompanied by changes in the macromolecular structure of the chromosome.	Activation of genes on polytene chromosomes is accompanied by "puffing."
RNA (post-transcriptional controls)	Primary transcripts may require processing to produce active cytoplasmic RNA species.	Ribosomal RNA gene transcripts are cleaved to yield rRNA molecules for ribosome assembly.

mRNA must be transported from the nucleus to the cytoplasm.	The drug cordycepin blocks appearance of new mRNA in the cytoplasm, probably by blocking synthesis of poly-A termini required for transport.
RNA degradation takes place both in the nucleus and in the cytoplasm.	Hemoglobin mRNA has a half-life of ∼100 days, whereas the average HeLa cell mRNA has a half-life of 2-3 hours.
mRNA in the cytoplasm may be "masked" in an inactive form.	Maternal mRNA in eggs remains inactive until after fertilization.
mRNA must attach to ribosomes for initiation of polypeptide synthesis.	A variety of initiation factors are involved in attachment of mRNA and initiation of polypeptide synthesis.
The post-translational fate of a polypeptide may depend on the class of ribosomes on which it is synthesized.	Polypeptides that are to be excreted or incorporated into the plasma membrane appear to be synthesized specifically on ribosomes that are bound to the membrane of the rough endoplasmic reticulum.
Translation on polyribosomes may occur at varying rates.	The α and β chains of hemoglobin appear to be translated at different rates.
Protein (post-translational controls)	
Proteins may be stored in vesicles and secreted when required.	Potentially destructive hydrolytic enzymes often are stored in lysosomes until needed.
Proteins may require modification of their covalent structure for activation.	Proinsulin must be cleaved to form active insulin; proline residues in procollagen must be hydroxylated and partially glycosylated to allow secretion and proper assembly of collagen fibers.
Proteins are degraded at different rates.	Antibodies of one class (IgG) have a half-life in the serum of 25 days, wheras antibodies of another class (IgE) have a half-life of 2 days.

probably at the DNA level, to code for a single polypeptide product. The genome of a mammal contains many different variable-region genes and only a few constant-region genes. It has been suggested that each antibody-producing cell becomes committed to the synthesis of a particular antibody by translocation and joining of a variable-region gene to a constant-region gene. This mechanism would be convenient for any system in which individual cells must differentiate to produce distinct proteins with constant as well as variable properties. Such proteins, for example, could be important as cell surface markers in cell-cell interaction during development. However, there is no evidence so far that a translocation mechanism is used outside of the immune system.

(c)   In a few lower organisms, such as the nematode worm *Ascaris,* development is accompanied by the loss of genetic material. In *Ascaris,* specific chromosomes are lost from somatic cells during the first five cleavage divisions, so that only the germ cell precursors retain a normal chromosomal complement. Loss of genetic material does not appear to occur during differentiation in higher animals.

5-8   Extensive regulation of gene expression occurs at the level of transcription

(a)   Although control at other levels occurs in eucaryotic cells, transcriptional control undoubtedly is a major mode of gene regulation. Evidence from two kinds of experiments supports this conclusion.

(1)   RNA-DNA association (hybridization) experiments (see Appendix to this chapter) show that different populations of nuclear RNA molecules are present in different developmental stages or different tissues of the same organism. These differences are seen in transcripts of both repetitive and unique DNA sequences (Table 5-2). For example, there is total lack of homology between repetitive-sequence transcripts from *Xenopus* oocytes and blastulas. Likewise, there is more than a 10-fold increase in complexity of unique-sequence transcripts during development of a mouse embryo.

(2)   Many attempts have been made to show that chromatin preparations from different stages or tissues of the same organism have different sequences available for transcription *in vitro* by RNA polymerase. Unfortunately there are at least two major pitfalls in this approach. First, since RNA polymerase binds nonspecifically to ends or single-strand breaks, even procaryotic DNA gives spurious transcripts *in vitro* unless isolated with extreme care to avoid breakage. It is even more difficult to know whether chromatin has been isolated intact, since its native state and physical properties are not yet understood as well as those of single DNA molecules. Second, there is no guarantee that the RNA polymerase used in these experiments, which generally is isolated from *E. coli,* will recognize the signals in

Table 5-2.   Sequence homology evidence for transcription level regulation. (Adapted from E. H. Davidson and R. J. Britten, *Quart. Rev. Biol.* **48,** 565, 1973.)

Cell types compared	Relative differences in RNA populations compared
Repetitive sequence transcripts	
Mouse L-cell, liver, spleen, kidney	15%-40%
Sea urchin eggs, blastula, gastrula	∿35%
Various stages of embryonic mouse liver	up to ∿50%
Xenopus oocytes and blastulae	total lack of homology (>90%)
Mouse liver and uterus ± estrogen stimulation	up to ∿25%
Nonrepetitive sequence transcripts	
Mouse liver, brain, and spleen	up to ∿30%
Mouse blastocyst and later stages (nuclear RNA only)	total complexity increases more than 10 fold
Mouse liver, spleen, and kidney	>70%
Chick oviduct ± estrogen stimulation (nuclear RNA only)	20%
Slime mold *(Dictyostelium)* at various stages of development	10%-40%

chromatin that promote or prevent transcription *in vivo*. For example, *in vitro* transcription of the 5S RNA genes of *Xenopus* chromatin with *E. coli* RNA polymerase gives preferential transcription of the wrong strand of DNA as well as the spacer region, which usually is not transcribed *in vivo*.

However, in other experiments, *in vitro* transcription of duck reticulocyte chromatin did produce RNA that included sequences characteristic of hemoglobin mRNA, whereas transcription of duck-liver chromatin did not. The assay for hemoglobin mRNA sequences was carried out by hybridization, using radioactive complementary single-stranded DNA that had been synthesized using RNA-directed DNA polymerase (reverse transcriptase), with purified hemoglobin mRNA as a template (Figure 5-8). Thus *in vitro* transcription of chromatin can yield significant results provided that sufficiently sophisticated assay techniques are employed.

(b)   Control of transcription may involve nonhistone chromosomal proteins that act as positive regulatory elements to turn on mRNA synthesis, although the evidence for this hypothesis is circumstantial.

(1)   Some of the nonhistone chromosomal proteins from rat-liver chromatin complex readily with rat DNA, less well with mouse DNA, and not at all with DNA from more distantly related mammals, suggesting some specificity for binding to homologous DNA.

(2)   Specific hormone-receptor proteins, which are likely to represent regulatory elements of some kind (see Essential Concept 6-8), are found among the nonhistone chromosomal proteins.

(3)   Cells that are actively expressing differentiated functions generally have higher levels of nonhistone chromosomal proteins than those that are undifferentiated or synthetically inactive. Moreover, the levels of nonhistone chromosomal proteins increase when a cell is stimulated experimentally to express a differentiated gene product. However, it is still unclear whether these differences represent causes or effects of altered mRNA synthesis.

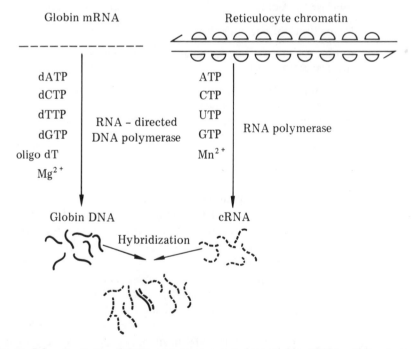

Figure 5-8. Synthesis of DNA complementary to hemoglobin mRNA and its use to detect synthesis of hemoglobin mRNA *in vitro*. (Adapted from R. Axel *et al., Cold Spring Harbor Symposia on Quantitative Biology* **38,** 773, 1973.)

5-9   Chromosome puffs allow visualization of temporally controlled transcription sequences

Activation of genes on polytene chromosomes of some developing insects can be seen in the light microscope as an apparent puffing of restricted chromosomal regions. When a particular gene is activated, the DNA in the corresponding chromosomal band appears to unfold as RNA synthesis is initiated. Reproducible sequences of puffing can be observed during development. On the salivary gland polytene chromosomes of early larval *Drosophila,* 10 bands form prominent puffs that are continuously active for about 30 hours. Ten hours before pupation a new puffing sequence begins suddenly in response to release of the steriod hormone ecdysone into the animal's hemolymph. During the next 12 hours, more than 125 new bands are activated. Each band puffs at a particular time after the initial hormonal signal, and is active for a characteristic period before regressing (Figure 5-9).

The same sequence of events can be reproduced *in vitro* with excised salivary glands incubated in the presence of ecdysone. On the basis of studies with this system, puffs that occur in the active 12-hour period can be divided into three classes: intermolt puffs, which are present before the hormonal signal; early puffs, which occur within a few minutes of hormone application; and late puffs, which arise several hours later. The early puffs appear to be activated by the hormone through a specific receptor protein (see Essential Concept 6-8). One or more proteins synthesized during the early puffing period, presumably as the result of synthesis and translation of new mRNA species, apparently promote the activation of late puffs. These observations directly support the conclusion that transcriptional control is an important mechanism for temporal regulation of gene expression in development.

5-10   Different RNA polymerases have specialized roles in eucaryotic transcription

(a)   Three types of RNA polymerase, designated I, II, and III, have been found in a variety of eucaryotic cells. These enzymes can be distinguished by their differential sensitivities to an inhibitor of RNA synthesis called *α-amanitin*. RNA polymerase I is not inhibited at all; RNA polymerase II is inhibited at low concentrations of the drug; and RNA polymerase III is inhibited only at high concentrations.

(b)   RNA polymerase I transcribes the genes for 18S and 28S rRNA. RNA polymerase II transcribes most other DNA sequences, and therefore presumably is responsible for mRNA synthesis. RNA polymerase III catalyzes the synthesis of tRNA and 5S ribosomal RNA. RNA polymerase III from mouse has been fractionated into two species, suggesting that there also may be separate enzymes for tRNA and 5S RNA synthesis.

(c)   The levels of RNA polymerases I and III seem to vary in parallel according to the physiological state of the cell. This suggests that the activity of the rRNA,

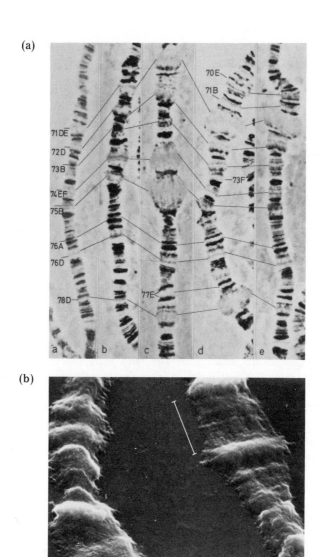

Figure 5-9. (a) Puffing sequences of a *Drosophila* chromosome during larval development. The numbers indicate specific chromosomal bands. [From M. Ashburner, "Puffing patterns in *Drosophila melanogaster* and related species," in W. Beermann (Ed.), *Developmental Studies on Giant Chromosomes,* Springer-Verlag, Berlin, 1972.] (b) Scanning electron micrograph of a segment of a polytene chromosome from *Drosophila*. The bar indicates a puff. (Courtesy Dr. H. Mitchell.)

tRNA, and 5S RNA genes might be regulated in part by the levels of these enzymes. In contrast, the levels of RNA polymerase II show much less variability.

(d)   The RNA polymerases are large proteins with complex subunit structures. For example, RNA polymerase II from calf thymus is composed of four polypeptide chains with approximate molecular weights of 190,000, 150,000, 35,000, and 25,000. The specific functions of these subunits, and the extent to which they are shared among the various polymerases are unknown.

5-11   Nuclear RNA transcripts fall into at least four classes

Transcription must result in the synthesis of precursors for mRNA, rRNA, and tRNA, which are modified if necessary and transported to the cytoplasm where they function. Unfortunately, many difficulties remain in reconciling these simple expectations with the observed features of nuclear transcripts and their apparent relationships to cytoplasmic RNA's. The nuclear RNA's can be divided somewhat arbitrarily into four classes, based on size, rate of turnover, sequence complexity, and function. The relative abundances of these classes vary with cell type and growth conditions.

(1)   The class termed heterogeneous nuclear RNA (hnRNA) is generally the most abundant. These molecules range from about 1000 to 50,000 nucleotides in length. Reassociation experiments show that in sea urchin embryos about 10% of the hnRNA sequences are repetitive, whereas the remaining sequences are unique. These repetitive and unique elements are known to be interspersed; that is, most hnRNA molecules include both kinds of sequences.

Pulse-chase labeling experiments show that only 5% to 20% of labeled hnRNA appears later in the cytoplasm as mRNA. The remainder apparently is degraded, and the nucleotides are reutilized for synthesis within the nucleus. This observation suggests two alternative possibilities regarding the relationship of hnRNA to mRNA. Either only a small fraction of the hnRNA molecules are mRNA precursors, or only a small fraction of each hnRNA molecule represents mRNA, the remainder being degraded before transport to the cytoplasm. The sequence complexity of hnRNA is about ten times that of cytoplasmic mRNA.

(2)   Transcripts of the genes for 18S and 28S rRNA, 5S rRNA, and tRNA are present in the nucleus as precursor molecules, which later become cleaved to produce the mature cytoplasmic forms of these RNA's. This cleavage is described in Essential Concept 5-12 for the precursor of 18S and 28S rRNA, the only RNA of this class that has been characterized in some detail.

(3)   Nuclei also contain low-molecular-weight RNA molecules of 100 to 300 nucleotides in length that constitute about 25% of the total nuclear RNA. This RNA is more stable than hnRNA, and its sequence complexity is low. Therefore, it cannot result from non-

selective degradation of the high-complexity hnRNA population. Due to its small size and low complexity, low-molecular-weight RNA is unlikely to represent mRNA precursors. The function of low-molecular-weight RNA in the nucleus is still unknown.

(4)   Another class of low-molecular-weight nuclear RNA has been isolated from chromatin and is called cRNA. It differs from the bulk low-molecular-weight RNA primarily in its higher sequence complexity, which has been shown by hybridization experiments to represent 5% to 10% of the middle-repetitive DNA sequences. The function of this RNA is also unknown.

5-12   Some nuclear RNA molecules are structurally altered before transport to the cytoplasm

It is not yet known whether all cytoplasmic RNA molecules undergo modifications of covalent structure in the course of their transport from the nucleus, but there is clear evidence that some such processing occurs.

(1)   rRNA is synthesized as a large 45S primary transcript that includes spacer sequences as well as the sequences of 18S and 28S rRNA. This transcript is processed in the nucleus, as shown in Figure 5-10. A series of cleavages yields 18S and 28S rRNA molecules, which then are transported to the cytoplasm for assembly into ribosomes.

(2)   Processing of a different sort has been suggested for mRNA precursors, based on the finding in cultured mammalian cells and sea urchin embryos that some hnRNA and many cytoplasmic mRNA molecules carry a sequence of about 200 adenylate (A) residues at their 3' ends. Complementary poly-T sequences of this length are not

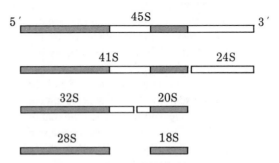

Figure 5-10. Processing of ribosomal RNA from HeLa cells. The shaded regions indicate sequences of mature 18S and 28S rRNA. (Adapted from P. K. Wellauer and I. B. Dawid, *Cold Spring Harbor Symposia on Quantitative Biology* **38,** 525, 1973.)

found in the DNA, and inhibitor experiments indicate that the poly-A sequences are added to hnRNA molecules after transcription is completed. Kinetics of labeling are consistent with the view that this processing occurs in the nucleus, and that poly-A-containing hnRNA molecules are transported to the cytoplasm to serve as mRNA. A functional role for poly-A in the transport process is suggested by experiments with the inhibitor 3'-deoxyadenosine (cordycepin), which preferentially blocks the synthesis of 3'-terminal poly-A sequences without affecting general hnRNA synthesis, and at the same time prevents the appearance of new mRNA molecules in the cytoplasm.

(3) Processing of mRNA by cleavage also has been suggested on the basis of differences between bulk hnRNA and cytoplasmic RNA. These two RNA populations show different, though overlapping, size distributions in higher eucaryotes. mRNA chain lengths range from about 1000 to 10,000 nucleotides, whereas hnRNA chain lengths appear to range from about 1000 to 50,000 nucleotides. In lower eucaryotes such as slime mold, *Amoeba,* and *Tetrahymena,* however, much smaller size differences are seen between the two RNA populations. In any case, since only a small fraction of hnRNA ever appears in the cytoplasm, comparisons of the total hnRNA and mRNA populations are inconclusive. Definitive evidence for processing by cleavage must be sought in experiments with specific individual mRNA molecules.

(4) In the fly *Chironomus,* very large puffs, called Balbiani rings, synthesize sufficiently large quantities of RNA that the products of a single activated locus can be isolated and analyzed. One of these puffs, designated BR2, has been studied in detail. BR2 appears to result from the unfolding of a single polytene band. The primary transcript of BR2 is a large RNA molecule between 15,000 and 30,000 nucleotides in length, which is transported to the cytoplasm with little if any modification in size. A protein coded for by this RNA has been tentatively identified as a polypeptide about 4400 residues in length, which would require a coding capacity of about 13,000 nucleotides in mRNA. These results suggest that at least some high-molecular-weight nuclear RNA's may represent atypically large mRNA molecules that code for large polypeptides.

(5) A few specific mRNA molecules have been isolated from poly-ribosomes precipitated by antibodies directed against the proteins coded by these messages. The two best studied examples are the mRNA's for hemoglobin from mammalian reticulocytes and ovalbumin from hen oviduct. Highly labeled complementary DNA molecules, synthesized *in vitro* using mRNA as a template for RNA-directed DNA polymerase (Figure 5-8), provide sensitive

hybridization probes for detection of the specific mRNA sequences in different RNA fractions. Using this approach, it has been established that both hemoglobin and ovalbumin mRNA's carry 3'-terminal poly-A sequences. However, the nuclear form of ovalbumin mRNA is the same size as the cytoplasmic mRNA, and the nuclear form of hemoglobin mRNA, while slightly larger than the cytoplasmic form, is considerably smaller than the average hnRNA molecule.

In summary, most mRNA precursors in the mRNA population may be processed by addition of poly-A sequences prior to transport from the nucleus. However, although many hnRNA molecules are larger than the largest cytoplasmic mRNA's, there is so far no evidence for mRNA processing that involves the extensive cleavage seen in rRNA processing. Clear answers on the extent and nature of mRNA processing must await studies on more examples of specific mRNA's and their nuclear precursors.

5-13   Eucaryotic mRNA's are translated on polyribosomes as monocistronic units

(a)   A major difference between procaryotic and eucaryotic protein synthesis is that cytoplasmic mRNA's in eucaryotes appear to carry information for only single polypeptides. All of the polysomes identified by immunoprecipitation as synthesizing specific proteins are close to the length predicted for the corresponding mRNA's. The size distribution of mRNA molecules in several cells is almost exactly as predicted from the size distribution of newly synthesized polypeptides (Figure 5-11). Moreover, double-labeling experiments on the ratio of nascent polypeptide to polynucleotide as a function of polysome size in several eucaryotic systems suggest that there are few if any polysomes that synthesize more than one protein.

Studies on the expression of viral genes in animal cells support these views. RNA viruses, which must synthesize several proteins using the synthetic machinery of the infected cell, apparently never present the cell with a polycistronic message. Some viruses carry their genome as a collection of small RNA fragments, whereas others produce fragments in the course of viral RNA replication. The genomes of still others, such as poliovirus, are translated intact as monocistronic messages to produce very large precursor polypeptides, which then are cleaved to form the various functional viral proteins.

(b)   Polyribosomes can be grouped into two functional classes that are located in different parts of the cell. Cytoplasmic polyribosomes, which are unattached to membranes, synthesize proteins destined for use in the cytoplasm or for the internal surface of the plasma membrane. Bound polyribosomes, which are attached to the membranes of the endoplasmic reticulum, synthesize proteins for the outer surface of the plasma membrane, as well as proteins that are to be packaged in vesicles for storage or secretion (WWBH Essential Concept 5-8).

(c) Evidence for control of gene expression at the translational level comes from the finding that the synthesis of mRNA often precedes its translation by a considerable time interval. Striking examples are found in sea urchin and amphibian development, where mRNA synthesized and incorporated into polyribosomes during oogenesis can remain dormant for several weeks until fertilization somehow triggers its translation.

(d) The molecular events of mRNA translation on ribosomes are similar in procaryotes and eucaryotes (WWBH Essential Concept 18-4). Initiation of the polypeptide involves an AUG codon and a unique species of methionyl-tRNA^{Met}, which is formylated in procaryotes but not in eucaryotes. Termination of the polypeptide in both kinds of cells occurs in response to the codons UAG, UAA, and UGA.

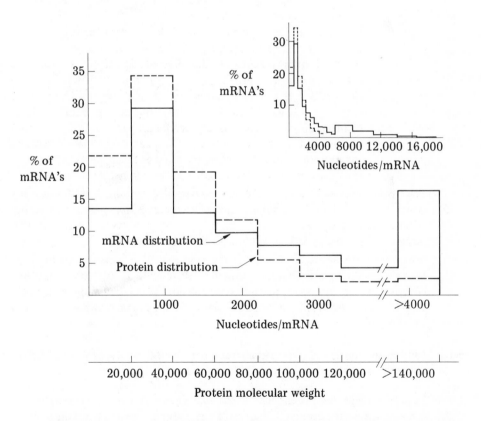

Figure 5-11. Size distributions of mRNA and polypeptides from HeLa cells. Solid line indicates lengths of mRNA molecules in nucleotides. Dashed line indicates the distribution of polypeptide molecular weights on a scale chosen so that the molecular weight of one amino acid residue corresponds to that of three nucleotides. Inset shows the complete size distribution of mRNA molecules. (Adapted from E. H. Davidson and R. J. Britten, *Quart. Rev. Biol.* **48,** 565, 1973.)

5-14   Post-translational modifications of proteins provide additional controls on gene expression

(a)   Following their synthesis on polyribosomes, many proteins require enzyme-catalyzed modifications of their primary structure to become active. Two common kinds of modifications are proteolytic cleavage and side-chain alterations.

(1)   Hydrolytic enzymes often are synthesized as inactive *proenzymes* (zymogens). An example is trypsinogen, the precursor of trypsin. Similarly, polypeptide hormones often are produced in the form of inactive *prohormones,* as exemplified by proinsulin, the precursor of insulin [see Essential Concept 6-2]. Activation of the inactive precursor forms depends upon removal of a specific polypeptide fragment by proteolytic cleavage.

(2)   Examples of side-chain modifications required for activity are found in the assembly of collagen fibers. The three polypeptide strands of procollagen, the precursor of the collagen subunit, will not fold into their characteristic triple helical structure (WWBH Essential Concept 4-7) until a large number of Pro and Lys side chains have been hydroxylated. Moreover, glycosylation of some of the resulting hydroxyl groups probably is required for proper subunit alignment in the subsequent assembly of fibers. More general examples are found among membrane proteins and mammalian serum proteins, many of which require glycosylation of specific side chains for activity.

(b)   The functions of some proteins are controlled by other specific proteins which inactivate them. Inactivation can be accomplished by association with a specific binding protein, such as the trypsin inhibitor found in mammalian serum, or by proteolytic degradation of the active protein.

(c)   The mechanisms considered in the two preceding sections involve irreversible modifications of primary structure. Reversible modifications that alter enzyme activity, such as the phosphorylation and dephosphorylation of enzymes in glycogen metabolism, are important in metabolic control (WWBH Essential Concept 9-4).

5-15   Models for control of gene expression must account for the process of differentiation and the organization of the genome

Several attempts have been made to formulate a useful working hypothesis that can account for the observed features of eucaryotic genes and point the way toward understanding how their expression is controlled, just as the Jacob-Monod operon model of the early 1960's helped to explain the control of procaryotic genes (Essential Concepts 1-3 through 1-5). The most comprehensive and commonly referred to of these attempts is a model first formulated in 1969 by Britten and Davidson. This model is designed to account for two distinctive features of complex eucaryotes: the temporally controlled expression of different sets of genes in different cell lines, and the organization of repetitive and unique sequences in the genome. The basic reasoning behind this model and its principal features are presented in the following sections.

(1)   The sequence of gene expression in any given cell line is postulated to result from a cascade of positive regulatory events at the transcriptional level, formally similar to those observed in procaryotes such as bacteriophage lambda (Essential Concept 1-9). In such a sequence, each activated group of structural genes produces one or more regulatory proteins that activate subsequent groups of structural genes. Initial differentiation in eucaryotes may be programmed when different blastula cells incorporate different regulatory proteins from the cytoplasm during cleavage of the fertilized egg [Essential Concept 5-3(a)]. These regulatory proteins in turn initiate different cascades of successive positive control in different cell lines. Subsequent differentiation also may depend on specific cytoplasmic inhomogeneities that result in distribution of newly synthesized regulatory proteins to only one of the two daughter cells upon cell division. The problem of differentiation thus becomes partly a problem of how regulatory proteins are placed in specific cytoplasmic locations.

(2)   The basic genetic elements required for sequential positive control are structural genes with adjacent control elements that are recognized by regulatory proteins, as well as the genes for these regulatory proteins. The control elements are called *receptor sequences* or *receptor genes* by Britten and Davidson, who propose that a structural gene is transcribed only when its adjacent receptor gene interacts with a specific regulatory protein. The set of functionally related structural genes that is activated by a particular regulatory protein is called a *battery* of structural genes.

(3)   The model, diagrammed in Figure 5-12, proposes that there are two classes of regulatory proteins, representing two levels of control. The primary regulatory proteins are called *sensor proteins,* since they may act only in response to *effectors,* such as hormones, in the local environment of a cell. The cytoplasmic steroid receptors found in the target cells of steroid hormones (Essential Concept 6-8) may be examples of sensor proteins.

Sensor proteins are postulated to act on an intermediate level of switching elements (sensor genes) that control the transcription of *integrator genes*. The integrator genes in turn code for the secondary regulatory proteins, called *activator proteins,* which interact with receptor sequences to promote structural gene expression.

Two kinds of evidence support the notion of intervening integrator genes rather than direct activation of structural genes by sensor proteins. First, recent findings suggest that the number of genes activated in response to a given signal may be considerably larger than the number of hormone receptor proteins in a target cell. Second, the properties of homoeotic mutants (Essential Concept 5-5) suggest that one developmental program can be switched to another by a single mutation. Such mutations are likely to represent alterations in switching elements. In terms of the model, they can be explained simply as translocations of sensor genes such that a sensor protein activates the wrong set of integrator genes.

(4)   Given such a model, two known properties of eucaryotes predict the existence of repetitive sequences in the genome. First, all the func-

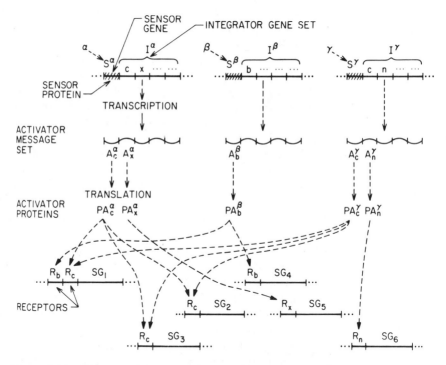

Figure 5-12. Elements of the Britten-Davidson model for gene control in eucaryotes. The diagram shows a simplified system in which three external effectors, α, β, and γ, control the expression of four gene batteries that include a total of six structural genes. Specific sensor proteins, S^α, S^β, and S^γ, when combined with the appropriate effector, activate the expression of integrator gene sets, which in turn code for specific activator proteins. Activator proteins are designated by a superscript indicating the integrator gene set in which they are encoded, and a subscript indicating the receptor sequence that they recognize (e.g., PA_c^α, PA_x^α, etc.). Activator proteins interact with receptor sequences (R_b, R_c, etc.) to activate the adjacent structural genes (SG_1, SG_3, etc.). Structural genes with the same receptor sequence constitute a battery. See text for additional explanation. (From E. H. Davidson and R. J. Britten, *Quart. Rev. Biol.* **48,** 565, 1973.)

tionally related genes in a battery are turned on by a single activator protein, and thus must share a common receptor gene. Generally, functionally related genes in eucaryotes are unlinked, in contrast to the situation in procaryotes. Therefore, common receptor sequences must be repeated the same number of times as there are genes in a battery. Second, batteries of structural genes must be overlapping sets, since many of the same proteins are required in different differentiated cells. Such overlap requires the same or similar activator proteins to be produced in response to different sensors. This requirement in turn demands that the same or similar integrator genes be repeated in a number of integrator gene sets, as many times as the number of batteries in which the corresponding structural gene is included. For example, in Figure 5-12 the structural genes SG_2 and SG_3, each controlled by the receptor sequence R_c, are included in two batteries of genes, turned on by sensors S^α and S^γ, respectively. Consequently, genes for activator proteins that recognize R_c must be included in both the S^α and S^γ integrator gene sets.

The Britten-Davidson model predicts that if most structural genes are composed of unique sequences and belong to functionally related sets whose members are unlinked, then much of the eucaryotic genome should consist of single-gene sized unique sequences interspersed with shorter repetitive receptor sequences. So far this expectation has been borne out by the observed patterns of sequence organization in several genomes (Essential Concept 2-12).

5-16   Additional concepts and techniques are presented in the Appendix and the Problems section

(a)   Analysis of RNA-DNA hybridization reactions. Appendix.

(b)   Estimating numbers of embryonic progenitor cells in experiments with allophenic mice. Problems 5-5 and 5-8.

(c)   Dual control of an enzyme level by a hormone and a substrate. Problem 5-12.

(d)   Apparent induction of enzymes by actinomycin D. Problem 5-14.

(e)   Analysis of RNA processing by mapping of intramolecular base pairing. Problem 5-15.

(f)   Processing of precursor proteins in animal virus infection. Problem 5-17.

(g)   The effect of 5-bromodeoxyuridine (BrdU) on expression of differentiated functions in cultured cells. Problem 5-20.

(h)   Estimation of mRNA complexities from the extents of DNA-driven hybridization reactions. Problem 5-21.

(i)   Estimation of the expressed fraction of the genome from RNA-driven hybridization experiments. Problem 5-23.

APPENDIX: ANALYSIS OF RNA-DNA HYBRIDIZATION REACTIONS

A5-1   Complementary RNA and DNA molecules can associate to form hybrid duplexes

(a)   RNA-DNA hybridization is a valuable technique for determining relationships between DNA sequences in the genome and the RNA products of their transcription. Two conditions for hybridization are considered in this Appendix: association reactions in which DNA is in large excess over RNA (DNA-excess reactions) and reactions in which RNA is in large excess over DNA (RNA-excess reactions). These two conditions yield different but complementary information.

(b)   RNA-DNA hybridization generally is carried out using the procedures described for DNA reassociation in Essential Concept A2-1. Either the RNA or the DNA is labeled radioactively to facilitate the analysis. Extents of hybridization reactions are measured by one of the following techniques.

> (1)   Under the appropriate conditions pancreatic ribonuclease (RNase) will catalyze hydrolysis of single-stranded RNA, but not of RNA in the form of an RNA-DNA hybrid duplex. Thus samples containing labeled RNA can be digested to completion and assayed for nuclease-resistant RNA.

> (2)   Hydroxyapatite chromatography can be used to separate hybrid duplexes from single-stranded RNA and DNA under partially denaturing conditions. Single-stranded RNA often binds to hydroxyapatite under conditions where single-stranded DNA does not (Essential Concept A2-1), due to the tendency of RNA to form intramolecular duplexes. However, in the presence of urea, formamide, or high concentrations of NaCl, single-stranded RNA does not bind, but duplex molecules still are retained. Under these conditions, association can be followed by passing samples through a hydroxyapatite column and determining the fraction of labeled RNA or DNA that is retained.

> (3)   Nitrocellulose filters retain DNA molecules, which aggregate and remain at the filter surface. RNA molecules in RNA-DNA duplexes are retained as well, whereas free RNA passes through the filter. Thus reaction mixtures containing labeled RNA can be analyzed by filtration and analysis of RNA remaining on the filter. Treatment of the filter with RNase before analysis removes uncomplexed portions of RNA molecules that are only partially in the duplex form.

(c)   Another convenient procedure for hybridization is carried out by immobilizing denatured DNA on nitrocellulose filters *prior* to the reaction and then incubating the filters with labeled RNA. Hybridization is measured as filter-bound radioactivity after RNase treatment and washing to remove unreacted RNA. Because the DNA molecules are not free in solution, this technique gives results that are quantitatively less reliable and more difficult to analyze kinetically than those of solution hybridization reactions. Filter hybridization therefore is not considered in the following discussion.

A5-2   DNA-excess reactions can be used to determine which frequency classes of DNA sequences are represented in an RNA population

(a)   In a DNA-excess association reaction the RNA is present in minute quantities as a radioactively labeled tracer. Two reactions are occurring under these conditions:

$$\text{DNA} + \text{DNA} \xrightarrow{k} \text{DNA:DNA}$$
$$\text{RNA} + \text{DNA} \xrightarrow{k} \text{RNA:DNA}$$

in which k (liters/mole second) is a second-order rate constant that varies with reaction conditions, as discussed in Essential Concept A2-2. To simplify the following derivations it is assumed that k is the same for the two association reactions, although this equivalence is only approximate in practice.* The rate of disappearance of single-stranded DNA is given by Equation A2-1,

$$-\frac{dC}{dt} = kC^2 \tag{A2-1}$$

in which C is the concentration of single-stranded DNA in moles of nucleotide per liter and t is time in seconds. As shown in Essential Concept A2-2, this expression can be integrated from initial conditions $t = 0$ and $C = C_0$ to yield Equation A2-2:

$$\frac{C}{C_0} = \frac{1}{1 + kC_0 t} \tag{A2-2}$$

or

$$C = \frac{C_0}{1 + kC_0 t} \tag{A5-1}$$

---

*For a more rigorous and detailed treatment of this subject, see E. H. Davidson *et al.*, *Cell,* in press, 1975.

The rate of disappearance of single-stranded RNA is given by an expression analogous to Equation A2-1:

$$-\frac{dR}{dt} = kRC \tag{A5-2}$$

in which R represents the concentration of single-stranded RNA. Substitution of Equation A5-1 into Equation A5-2 gives

$$-\frac{dR}{dt} = kR \left(\frac{C_0}{1 + kC_0 t} \right) \tag{A5-3}$$

or

$$\frac{dR}{R} = -kC_0 \left(\frac{1}{1 + kC_0 t} \right) dt$$

This expression can be integrated from initial conditions $t = 0$ and $R = R_0$ to yield

$$\ln \frac{R}{R_0} = \ln \frac{1}{1 + kC_0 t}$$

or

$$\frac{R}{R_0} = \frac{1}{1 + kC_0 t} \tag{A5-4}$$

A comparison of Equations A5-4 and A2-2 indicates that the RNA sequences in a DNA-excess reaction associate at the same rate as the DNA sequences. Since the rate of association depends only upon the initial DNA concentration, C_0, hybridization reactions with excess DNA often are called *DNA-driven* reactions.

(b)   Since RNA tracer sequences associate with the same kinetics as the corresponding DNA sequences, the $C_0 t_{\frac{1}{2}}$ for a species of RNA indicates the frequency class of the DNA sequence from which the RNA was transcribed. As an example, consider the hypothetical eucaryotic genome discussed in Essential Concept A2-4. One half of this genome consists of sequences with a total complexity (X) of 10^5 nucleotide pairs, each repeated 10^4 times, and the other half consists of unique sequences with a total complexity of 10^9 nucleotide pairs.

Analysis of DNA reassociation kinetics gave $C_0t_{\frac{1}{2}}$ (mixture) values for these components of 2×10^{-1} mole sec/liter and 2×10^3 mole sec/liter, respectively. If radioactively labeled RNA from embryonic cells of this organism were isolated and used to perform a DNA-excess hybridization reaction, the results shown in Figure 5-13 might be found. Reassociation of the bulk DNA is followed by hypochromicity (Essential Concept A2-1), and hybridization of the labeled RNA is monitored by analysis of RNase-resistant radioactivity. The results indicate that 25% of the RNA sequences are transcribed from repeated DNA sequences, and 75% of the RNA sequences are transcribed from unique DNA sequences.

(c) Using a combination of hydroxyapatite chromatography and RNase resistance to analyze hybrid formation, it is possible to distinguish between RNA molecules that are transcribed completely from one DNA frequency class and molecules that contain sequences of two classes, as the result of transcription from adjacent repetitive and unique sequences in the genome. For example, analysis of the hybridization reaction of the preceding paragraph by these two techniques might give the results shown in Figure 5-14, which indicates that although only 25% of the RNA *sequences* are transcribed from repetitive DNA, 100% of the RNA *molecules* include sequences transcribed from repetitive DNA.

(d)   As mentioned in Essential Concepts 5-8(a) and 5-11, the properties of specific RNA transcripts often can be studied most conveniently by synthesis of a highly labeled complementary DNA sequence (cDNA) with RNA-directed DNA polymerase, followed by association analysis of the cDNA. For experimental purposes, a cDNA tracer behaves identically to an RNA tracer in a DNA-driven association reaction, so that the frequency class of a cDNA species can be determined from its $C_0t_{\frac{1}{2}}$ using hydroxyapatite chromatography as in the preceding example.

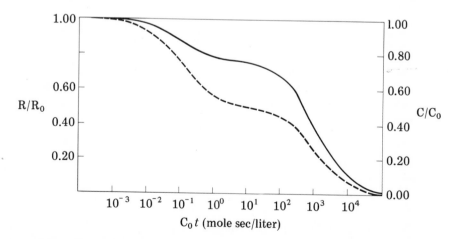

Figure 5-13. Association kinetics of tracer RNA (solid line) and bulk DNA (dashed line) in a hypothetical DNA-driven hybridization reaction.

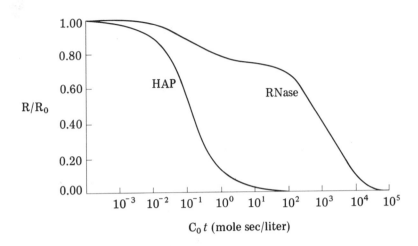

Figure 5-14. Analysis of the RNA hybridization in Figure 5-13 by two methods. The hydroxyapatite column retains RNA molecules that are hybridized to DNA over any portion of their sequence. The RNase assay measures only sequences that actually base-pair to DNA.

A5-3   RNA-excess hybridization can be used to measure directly the complexity of an RNA population

(a)   In an RNA-excess or *RNA-driven* association reaction, the following two reactions occur:

$$RNA + DNA \xrightarrow{k} RNA{:}DNA$$
$$DNA + DNA \xrightarrow{k} DNA{:}DNA$$

However, since the DNA is present in very low concentration as a radioactive tracer, the DNA-DNA reaction usually can be neglected. Under these conditions, the rate of disappearance of single-stranded DNA is given by the equation

$$-\frac{dC}{dt} = kRC \qquad\qquad (A5\text{-}5)$$

Since the RNA is present in large excess over DNA, the concentration of single-stranded RNA will not change appreciably during the reaction, that is, $R = R_0$. Thus Equation A5-5 becomes

$$-\frac{dC}{dt} = k\mathrm{R}_0 C$$

or

$$\frac{dC}{C} = -k\mathrm{R}_0 dt \tag{A5-6}$$

Integration of this expression from initial conditions $t = 0$ and $C = C_0$ yields

$$\ln \frac{C}{C_0} = -k\mathrm{R}_0 t$$

or

$$\frac{C}{C_0} = e^{-k\mathrm{R}_0 t}$$

Thus an RNA-driven hybridization is a pseudo first-order reaction, in contrast to the second-order association reactions discussed so far (Figure 5-15). The value of $\mathrm{R}_0 t$ when the reaction is half complete ($\mathrm{R}_0 t_{\frac{1}{2}}$) is given by

$$\mathrm{R}_0 t_{1/2} = \frac{\ln 2}{k} \tag{A5-7}$$

(b)   The sequence complexity of an RNA population can be determined from either the $\mathrm{R}_0 t_{\frac{1}{2}}$ or the end point of an RNA-driven association reaction.

    (1)   In a DNA reassociation reaction, the sequence complexity (X) is related to $C_0 t_{\frac{1}{2}}$ by a proportionality constant, K, whose value depends upon the reaction conditions (Essential Concept A2-3):

$$X = KC_0 t_{1/2} \tag{A2-3}$$

        The rate constant for DNA-DNA reassociation is related to $C_0 t_{\frac{1}{2}}$ by the expression

$$k = \frac{1}{C_0 t_{1/2}} \tag{A5-8}$$

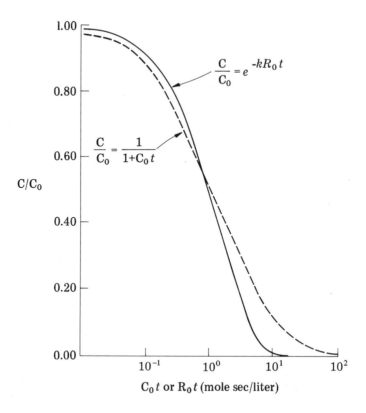

Figure 5-15.   Kinetics of single-stranded DNA disappearance in pseudo first-order and second-order reactions. Solid line represents C/C_0 as a function of R_0t in an ideal pseudo first-order RNA-driven hybridization reaction in which the DNA tracer contains sequences of only one frequency class, all of which are present in the RNA population. Dashed line represents C/C_0 as a function of C_0t in an ideal second-order DNA reassociation reaction.

(Essential Concept A2-2); the rate constant for RNA-DNA association is related to $R_0t_{\frac{1}{2}}$ by Equation A5-7. Under equivalent conditions, assuming equal rate constants, these two expressions can be equated if a factor of two is included to account for the fact that for a given set of sequences, the concentration of nucleic acid molecules from a single-stranded RNA preparation will be only half the concentration of the molecules from a double-stranded DNA preparation. Accordingly, Equation A5-7 must be modified to

$$R_0t_{1/2} = \frac{1}{2}\left(\frac{\ln 2}{k}\right)$$

or

$$k = \frac{\ln 2}{2R_0 t_{1/2}} \qquad (A5\text{-}9)$$

Then, equating the two expressions for k,

$$\frac{1}{C_0 t_{1/2}} = \frac{\ln 2}{2R_0 t_{1/2}}$$

or

$$C_0 t_{1/2} = \frac{2R_0 t_{1/2}}{\ln 2} \qquad (A5\text{-}10)$$

Substitution of Equation A5-10 into Equation A2-3 gives a relationship between RNA complexity and $R_0 t_{\frac{1}{2}}$:

$$X = \frac{2K R_0 t_{1/2}}{\ln 2} \qquad (A5\text{-}11)$$

Under standard association conditions, $K \simeq 5 \times 10^5$ liter nucleotides/mole second (Essential Concept A2-3).

(2) RNA sequence complexity can be determined more directly from the amount of tracer DNA that has associated to form hybrid duplexes at the end of an RNA-driven hybridization reaction. To simplify interpretation, such experiments generally are carried out using an isolated DNA frequency component (e.g., unique DNA from a eucaryotic genome) as the DNA tracer. At the completion of the reaction, the fraction of DNA that has associated with RNA indicates the fraction of the DNA sequences that are present in the RNA population. If the complexity of the tracer DNA is known, the complexity of the RNA population can be calculated from the extent of reaction.

As an example, consider an experiment in which a large excess of mRNA from cells of the hypothetical eucaryote of Essential Concept A5-2 is associated in an RNA-driven reaction with radioactive tracer DNA representing the unique fraction of the genome ($X = 10^9$ nucleotide pairs). A possible result is shown in

Figure 5-16. Since mRNA is transcribed asymmetrically, that is, from only one of the two DNA strands for any given sequence, then a maximum of 50% of the DNA could hybridize if the complexities of the mRNA (in nucleotides) and the DNA (in nucleotide pairs) were the same. Thus the 1% of DNA hybridized at the end of the reaction represents 2% of the DNA complexity, and the complexity of the RNA population is at least $0.02 \times 10^9 = 2 \times 10^7$ nucleotides. This figure is a minimum value, because the RNA population also may contain sequences that are not present in the tracer DNA from the unique fraction. This possibility can be checked using the kinetic approach described in the preceding paragraph.

A5-4 Homologies between RNA populations can be determined by hybridization competition under DNA-excess conditions

In a DNA-driven hybridization reaction with radioactive RNA, formation of labeled hybrids will be prevented by the addition of sufficient unlabeled RNA that contains sequences in common with the labeled tracer. This technique, called *hybridization competition,* can be used to measure the relatedness of two RNA populations, for example, labeled tracer RNA from one cell type and

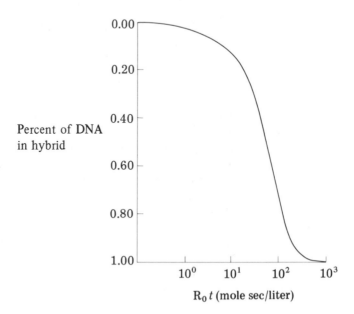

Figure 5-16. Determination of RNA sequence complexity from the final extent of tracer DNA association in an RNA-driven hybridization reaction. Hybrid is measured as radioactive tracer DNA retained by a hydroxyapatite column.

unlabeled competitor mRNA from another cell type. The fraction of the tracer RNA that can be prevented from hybridizing by increasing amounts of the competitor indicates the fraction of the tracer sequences that also are present in the competitor RNA population. For some examples of this technique, see WWBH Essential Concept 18-6.

REFERENCES

Where to begin

E. Hadorn, "Transdetermination in cells," *Scientific American* (November, 1968)

O. L. Miller, Jr., "The visualization of genes in action," *Scientific American* (March, 1973)

J. D. Watson, *Molecular Biology of the Gene* Chapter 16 (W. A. Benjamin, Menlo Park, Calif., 1970, 2nd ed.)

General

E. H. Davidson and R. J. Britten, "Organization, transcription and regulation in the animal genome," *Quart. Rev. Biol.* **48,** 565 (1973)

J. B. Gurdon, *The Control of Gene Expression in Animal Development* (Oxford University Press, Oxford and Harvard University Press, Cambridge, Mass., 1974)

B. Mintz, "Genetic control of mammalian differentiation," *Ann. Rev. Genetics* **8,** 411 (1974)

Chromosome Structure and Function: Cold Spring Harbor Symposia on Quantitative Biology, Vol. 38 (Cold Spring Harbor Laboratory, Cold Spring Harbor, N. Y., 1973)

Determination in early development

E. H. Davidson, *Gene Activity in Early Development* (Academic Press, New York, 1968)

A. Garen and W. Gehring, "Repair of the lethal development defect in *deep orange* embryos of *Drosophila* by injection of normal egg cytoplasm," *Proc. Natl. Acad. Sci. (U.S.)* **69,** 2982 (1972)

W. Gehring, "Genetic control of determination in the *Drosophila* embryo," in F. H. Ruddle (Ed.), *Genetic Mechanisms of Development,* 31st Symposium of the Society for Developmental Biology, p. 103 (Academic Press, New York, 1973)

B. Mintz, "Gene expression in allophenic mice," *Symposia of the International Society for Cell Biology* **9,** 15 (1970)

J. H. Postlethwait and H. Schneiderman, "Developmental genetics of *Drosophila* imaginal discs," *Ann. Rev. Genetics* **7,** 381 (1973)

L. D. Smith and R. E. Ecker, "Regulatory process in the maturation and early cleavage of amphibian eggs," *Current Topics in Develop. Biol.* **5,** 1 (1970)

D. Wimber and D. Steffensen, "Localization of gene function," *Ann. Rev. Genetics* **7,** 205 (1973)

Regulation of gene expression

R. Axel, H. Cedar, and G. Felsenfeld, "Synthesis of globin ribonucleic acid from duck-reticulocyte chromatin *in vitro,*" *Proc. Natl. Acad. Sci. (U.S.)* **70,** 2029 (1973)

J. D. Baxter, G. G. Rousseau, M. C. Benson, R. L. Garcea, J. Ito, and G. M. Tomkins, "Role of DNA and specific cytoplasmic receptors in glucocorticoid action," *Proc. Natl. Acad. Sci. (U.S.)* **69,** 1892 (1972)

W. Beermann (Ed.), *Developmental Studies on Giant Chromosomes* (Springer-Verlag, New York and Berlin, 1972)

J. O. Bishop, J. G. Morton, M. Rosbash, and M. Richardson, "Three abundance classes in HeLa cell messenger RNA," *Nature* **250,** 199 (1974)

G. Brawerman, "Eukaryotic messenger RNA," *Ann. Rev. Biochem.* **43,** 621 (1974)

B. Daneholt and H. Hosick, "The transcription unit in Balbiani ring 2 of *Chironomus tentans,*" *Cold Spring Harbor Symposia on Quantitative Biology* **38,** 629 (1973)

J. Darnell, W. Jelinek, and G. Molloy, "Biogenesis of mRNA: genetic regulation in mammalian cells," *Science* **181,** 1215 (1973)

A. L. Goldberg and J. F. Dice, "Intracellular protein degradation in mammalian and bacterial cells," *Ann. Rev. Biochem.* **43,** 835 (1974)

R. Goldberg, G. Galau, R. Britten, and E. Davidson, "Nonrepetitive DNA sequence representation in sea urchin embryo messenger RNA," *Proc. Natl. Acad. Sci. (U.S.)* **70,** 3516 (1973)

U. Grossbach, "Chromosome puffs and gene expression in polytene cells," *Cold Spring Harbor Symposia on Quantitative Biology* **38,** 619 (1973)

J. B. Gurdon, C. Lane, H. Woodland, and G. Marbaix, "Use of frog eggs and oocytes for the study of messenger RNA and its translation in living cells," *Nature* **233,** 177 (1971)

W. E. Hahn and C. D. Laird, "Transcription of nonrepeated DNA in mouse brain," *Science* **173,** 158 (1971)

B. Hamkalo and O. Miller, "Electronmicroscopy of genetic activity," *Ann. Rev. Biochem.* **42,** 379 (1973)

H. Holtzer, H. Weintraub, R. Mayne, and B. Mochan, "The cell cycle, cell lineages and cell differentiation," *Current Topics in Develop. Biol.* **7,** 229 (1972)

L. Hood, "Two genes-one polypeptide: fact or fiction?" *Federation Proceedings* **31,** 177 (1972)

D. Hourcade, D. Dressler, and J. Wolfson, "The nucleolus and the rolling circle," *Cold Spring Harbor Symposia on Quantitative Biology* **38,** 537 (1973)

D. Kabat, "Gene selection in hemoglobin and in antibody-synthesizing cells," *Science* **175,** 134 (1972)

F. Kafatos, "The cocoonase zymogen cells of silk moths: a model of terminal cell differentiation for specific protein synthesis," *Current Topics in Develop. Biol.* **7,** 125 (1972)

P. Leder, J. Ross, J. Gielen, S. Packman, Y. Ikawa, H. Aviv, and D. Swan, "Regulated expression of mammalian genes: globin and immunoglobulin as model systems," *Cold Spring Harbor Symposia on Quantitative Biology* **38,** 753 (1973)

H. Lodish, R. Firtel, and A. Jacobson, "Transcription and structure of the genome of the cellular slime mold *Dictyostelium discoideum,*" *Cold Spring Harbor Symposia on Quantitative Biology* **38,** 899 (1973)

P. A. Marks and R. A. Rifkind, "Protein synthesis: its control in erythropoiesis," *Science* **175,** 955 (1972)

G. S. McKnight and R. T. Schimke, "Ovalbumin messenger RNA: Evidence that the initial product of transcription is the same size as polysomal ovalbumin messenger," *Proc. Natl. Acad. Sci. (U.S.)* **71,** 4329 (1974)

G. S. Stein, T. C. Spelsberg, and L. J. Kleinsmith, "Nonhistone chromosomal proteins and gene regulation," *Science* **183,** 817 (1974)

D. F. Steiner, W. Kemmler, H. S. Tager, and J. D. Peterson, "Proteolytic processing in the biosynthesis of insulin and other proteins," *Federation Proceedings* **33,** 2105 (1974)

The Britten-Davidson model for control of gene expression

R. J. Britten and E. H. Davidson, "Gene regulation for higher cells: a theory," *Science* **165,** 349 (1969)

R. J. Britten and E. H. Davidson, "Repetitive and non-repetitive DNA sequences and a speculation on the origins of evolutionary novelty," *Quart. Rev. Biol.* **46,** 111 (1971)

Appendix: RNA-DNA hybridization reactions

E. H. Davidson, B. R. Hough, W. H. Klein, and R. J. Britten, "Structural genes adjacent to interspersed repetitive DNA sequences," *Cell* **4,** 217 (1975)

G. A. Galau, R. J. Britten, and E. H. Davidson, "Measurement of the sequence complexity of polysomal messenger RNA in sea urchin embryos," *Cell* **2,** 9 (1974)

PROBLEMS

5-1   Answer the following with true or false. If false explain why.

(a)   All the information for constructing a new organism probably is contained in each somatic cell.

(b)   In the developmental pathway of a particular cell, determination occurs earlier than differentiation.

(c)   An undetermined cell differentiates to become a specific cell type in a single step.

(d)   The stem cell for erythrocytes is a differentiated cell.

(e)   In embryos of many species the developmental fate of each blastomere is determined at least partially by the portion of egg cytoplasm that it receives.

(f)   Each tissue in a higher organism is derived from a single progenitor cell.

(g)   Male and female cells can be found with equal probability in any tissue of a *Drosophila* gynandromorph.

(h)   Shortly after emergence from the pupa, one kind of mutant fly, "wings-up," raises its wings perpendicular to its body axis and keeps them permanently in that position. Study of gynandromorphs that are heterozygous for this mutation shows that the character is associated more closely with the thorax of the fly than with the head or abdomen. From such studies one can conclude that the wings-up gene probably does not function at the level of the central nervous system.

(i)   Allophenic mice are twice as large as normal, because they are produced from two aggregated embryos.

(j)   Allophenic mice have a normal diploid karyotype.

(k)   Isocitrate dehydrogenase from two different inbred strains of mice differs in electrophoretic mobility. Liver cells from allophenic mice made from these two inbred strains show only the two parental electrophoretic forms. Since no intermediate hybrid enzyme forms are observed, isocitrate dehydrogenase must have a single subunit.

(l)   It is possible to construct a fate map of primordial stem cells in the embryo by studying *Drosophila* gynandromorphs, because the orientation of the first nuclear division is random with respect to the axis of the egg.

(m)   Imaginal discs normally are programmed to switch from one determined state to another.

(n)   A wing imaginal disc that is transplanted into the abdomen of an adult fly will metamorphose rapidly into a new wing structure.

(o)   Post-transcriptional controls of gene expression may include mRNA processing, transport to the cytoplasm, and degradation.

(p)   Specific RNA polymerases exist for most genes in eucaryotes.

(q)   *E. coli* RNA polymerase faithfully transcribes all eucaryotic genes.

(r)   cDNA transcripts of purified eucaryotic mRNA are useful for probing the control of gene expression because they can be labeled radioactively to a much higher specific activity than mRNA.

(s)   Most hnRNA and mRNA molecules carry segments of poly-A attached to their 3' ends.

(t)   In the course of early pupal development the insect hormone ecdysone directly causes more than 125 chromosomal bands to become puffs.

(u)   hnRNA, cRNA, and low-molecular-weight RNA are unique to eucaryotes.

(v)   The Britten-Davidson model for gene regulation at the transcriptional level assumes that genes are inactive unless specifically activated.

(w)   mRNA in procaryotes and eucaryotes is almost entirely monocistronic.

(x)   Membrane-bound ribosomes synthesize cytoplasmic proteins predominantly.

5-2   (a)   _____ is the process by which a cell becomes committed to the future expression of a specific phenotype.

(b)   Expression of a predetermined phenotype is called _____.

(c)   Experiments with _____ mice have shown that several cells in the embryo can be determined independently and identically.

(d)   Most structures of an adult *Drosophila* arise from _____ in the larva.

(e)   The process by which an imaginal disc changes from one determined state to another is called _____.

(f)   In _____ mutants of *Drosophila* one body part is replaced by another.

(g)   _____ is heterogeneous in size and turns over rapidly in the nucleus.

(h)   A thousand or more copies of ribosomal genes are produced in certain oocytes by _____ .

(i)   _____ ribosomes synthesize proteins that are secreted from the cell.

(j)   RNA polymerase II is inhibited by a low concentration of _____ .

(k)   _____ proteins may act as positive regulatory elements to turn on mRNA synthesis.

(l)   _____ transcribes the genes for 18S and 28S rRNA.

(m)   In the Britten-Davidson model of gene regulation, the external signals that initiate new developmental pathways are called _____ molecules.

5-3 Why are histones not considered likely candidates for the primary regulators of gene expression?

5-4 Indicate whether the following cells are differentiated or determined.
(a) Red blood cell
(b) Fibroblast
(c) Sperm
(d) A cell from a wing imaginal disc
(e) Reticulocyte
(f) Erythroid stem cell
(g) Hematopoietic stem cell
(h) A leg imaginal disc cell that has been transdetermined to a wing disc cell

5-5 Allophenic mice are produced by mixing the blastomeres from two different inbred strains that each express one of two allelic forms of an enzyme, Y^a or Y^b. When expression of the enzyme is examined in the granulocytes, megakaryocytes, and erythrocytes (three kinds of blood cells) of such mice, the data shown in Table 5-3 are obtained.
(a) How many progenitor cells are there for each cell type? How many for the entire system?
(b) What would be the probability of obtaining these results if each cell type were derived from two progenitor cells? If the entire system were derived from two progenitor cells?

Table 5-3. Expression of alleles of enzyme Y in the blood cells of allophenic mice (Problem 5-5).

Individual mouse	Granulocytes	Megakaryocytes	Erythrocytes
1	Y^a	Y^a	Y^a
2	Y^a	Y^a	Y^a
3	Y^b	Y^b	Y^b
4	Y^a	Y^a	Y^a
5	Y^b	Y^b	Y^b
6	Y^a	Y^a	Y^a
7	Y^b	Y^b	Y^b
8	Y^b	Y^b	Y^b
9	Y^b	Y^b	Y^b
10	Y^a	Y^a	Y^a
11	Y^b	Y^b	Y^b
12	Y^a	Y^a	Y^a
13	Y^a	Y^a	Y^a
14	Y^b	Y^b	Y^b
15	Y^b	Y^b	Y^b

5-6 Gynandromorph analysis of *Drosophila* is performed on flies that carry a single copy of the mutation of interest and a recessive visible marker, such as body color, both on the normal (nonring) X chromosome. Cells that lose the ring-X chromosome, which carries the wild-type alleles, express the mutant phenotype and can be identified readily by the marker phenotype.

One kind of *Drosophila* behavioral mutant, called "drop-dead," begins adult life as an apparently normal fly. However, at some unpredictable time, each individual becomes less active, begins to walk in an uncoordinated manner, falls on its back with limbs twitching, and dies. Gynandromorph analysis shows that this trait is more closely associated with the head than with the thorax or abdomen. However, occasional gynandromorphs whose entire head surface is wild-type drop dead. If you assume that the trait is associated with the head, how can you explain this observation?

5-7 Two inbred strains of mice, A and B, produce electrophoretically distinguishable isocitrate dehydrogenases (isozymes). Electrophoretic patterns for isocitrate dehydrogenase from various mice are given in Figure 5-17.
(a) How many subunits does isocitrate dehydrogenase have?
(b) Why does skeletal muscle differ from other tissues in allophenic mice?

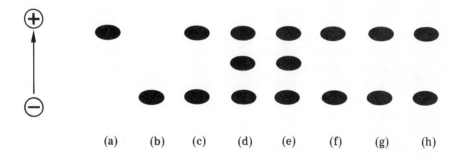

Figure 5-17. Patterns obtained by electrophoresis of isocitrate dehydrogenase from homogenates of various mouse tissues (Problem 5-7). (a) Pooled tissue from strain A; (b) pooled tissue from strain B; (c) pooled tissue from strains A and B; (d) pooled tissue from an F1 heterozygote obtained by mating strains A and B. Patterns (e)-(g) represent samples from various tissues of allophenic mice produced by combining embryos of strains A and B: (e) skeletal muscle; (f) liver; and (g) cardiac muscle. The pattern shown in (h) was obtained in a control electrophoresis of pure A enzyme and pure B enzyme.

5-8 The word "embryo" for the earliest stages of mammalian development is something of a misnomer, since only some cells of the inner cell mass in the blastocyst form the embryo proper, while the remaining cells become extraembryonic tissues and placenta. When blastomeres with different genotypes are aggregated to form single embryos, some of the resulting animals are true genetic mosaics, but others have only one or the other cellular genotype. The frequency of nonmosaics can be used as a clue to the determination of embryo versus extraembryonic cells. In one group of 129 allophenic mice analyzed for mosaicism, 75% were mosaics. Assuming that cells of each genotype have an equal chance of being determined as progenitor cells for the embryo, what is the most probable number of embryonic progenitor cells determined in the blastocyst?

5-9　(a)　An amphibian egg produces and stores about 10^{12} ribosomes in preparation for the tremendous rate of protein synthesis that accompanies the rapid cell cleavages during early development. How long would it take an egg to synthesize 10^{12} ribosomes at a rate typical of somatic cells (3×10^6 ribosomes/day)?
(b)　How does an amphibian egg increase its rate of ribosome production?
(c)　In four days a silk gland cell can synthesize 10^{10} silk fibroin molecules from two copies of the silk fibroin gene. Can you rationalize this observation in view of your calculations in Parts a and b?

5-10　Hybridization experiments using labeled 28S rRNA and unlabeled DNA from *Xenopus laevis* demonstrate that at saturation, 0.070 μg of 28S rRNA (molecular weight 1.6×10^6) is bound for each 100 μg of DNA. *Xenopus* somatic cells contain 3.6×10^{12} daltons of DNA. Calculate the number of 28S rRNA genes per diploid genome of *Xenopus*.

5-11　Upon digestion with T1 ribonuclease (cleaves on the 3′ side of G residues) mammalian hnRNA and mRNA yield long poly-A segments that contain no G residues. Upon digestion with pancreatic ribonuclease (cleaves on the 3′ side of U and C residues) they yield long poly-A segments that contain no U or C residues. What is the location of the poly-A segments in mammalian hnRNA and mRNA molecules?

5-12　HeLa cell chromatin contains 1.08 mg of histone and 0.7 mg of nonhistone chromosomal protein per mg of DNA. Assume that 10^6 daltons of DNA correspond to an average gene and that the average molecular weights of histones and of nonhistone chromosomal proteins are 12,000 and 70,000, respectively.
(a)　How many molecules of chromosomal protein are complexed with an average gene?
(b)　If an average HeLa cell contains 10^{13} daltons of DNA, how many molecules of chromosomal proteins are complexed with it?
(c)　If there are specific activator proteins for different batteries of structural genes, as proposed by Britten and Davidson, then these activators must be included among the chromosomal proteins. The feasibility of isolating activator proteins for detailed study will depend in part upon their relative abundance in chromatin. Consider a specific activator protein in HeLa cells that activates a battery of 1000 structural genes. Assuming that one molecule of activator is required per gene, and that the cells are diploid, what numerical percentage of the total chromosomal proteins would this protein represent?

5-13   The enzyme tryptophan pyrrolase catalyzes cleavage of the indole ring of tryptophan as a first step in the synthesis of an essential vitamin, nicotinamide. The total activity of tryptophan pyrrolase in rat-liver cells is increased by injection of tryptophan or the hormone hydrocortisone, as shown in Figure 5-18. In general, an increase in enzyme activity can result from an increase in its rate of synthesis, a decrease in its rate of degradation, or an increase in its catalytic efficiency. The catalytic parameters of tryptophan pyrrolase from treated and untreated rats are known to be the same. Suppose that you carry out two experiments to detect changes in rates of synthesis or degradation.

I.   You inject rats with saline solution (as a control), hydrocortisone, or tryptophan. After three hours you inject radioactive amino acids. A short time later you remove the livers and purify tryptophan pyrrolase. The specific radioactivity of the enzyme (cpm/μg) after each treatment is shown in Table 5-4.

II.   You inject rats with radioactive amino acids. After they have been incorporated into protein (no free radioactive amino acids remaining), you inject

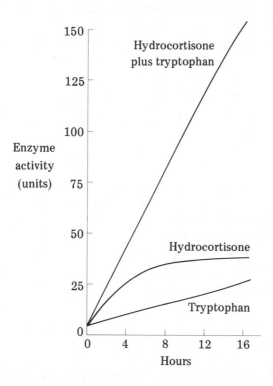

Figure 5-18.   Increase in tryptophan pyrrolase activity produced by administration of hydrocortisone and/or tryptophan (Problem 5-13). (Adapted from R. T. Schimke, *National Cancer Institute Monograph* **27,** 301, 1967.)

Table 5-4. Effects of tryptophan and hydrocortisone on synthesis of tryptophan pyrrolase (Problem 5-13). (Data from R. T. Schimke, *National Cancer Institute Monograph* **27**, 301, 1967.)

Injected solution	Specific radioactivity of enzyme (arbitrary units)
Saline	1.0
Tryptophan	1.4
Hydrocortisone	6.7

saline solution, or tryptophan. You then determine the amount of labeled tryptophan pyrrolase present at various times after injection as indicated in Figure 5-19.

(a) How do tryptophan and hydrocortisone increase the activity of tryptophan pyrrolase?

(b) Why are the effects of tryptophan and hydrocortisone not simply additive in Figure 5-18?

(c) Would you expect actinomycin D to alter the results shown in Figure 5-18?

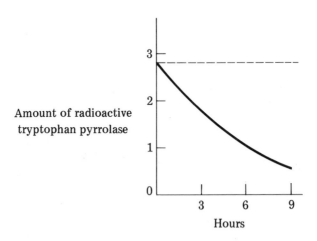

Figure 5-19. Effects of tryptophan on tryptophan pyrrolase stability (Problem 5-13). Prior to the experiment, rats were allowed to incorporate radioactively labeled amino acids into protein. At time 0, the rats were injected either with saline (solid line), or tryptophan (dashed line), and sacrificed at 3-hour intervals thereafter for determination of radioactivity remaining in tryptophan pyrrolase. (Adapted from R. T. Schimke, *National Cancer Institute Monograph* **27**, 301, 1967.)

5-14 Many researchers who have tested the effect of actinomycin D on the activity of a specific enzyme in eucaryotic cells have been surprised by an anomalous *increase* in total enzyme activity. This phenomenon is shown for tyrosine aminotransferase in rat hepatoma cells in Figure 5-20(a). Suppose that you have prepared antibody specific for tyrosine aminotransferase, so you rapidly can purify the enzyme to near homogeneity by immunoprecipitation. You determine the rates of synthesis and degradation of tyrosine aminotransferase in the presence and absence of actinomycin D by assaying, respectively, the uptake rate of radioactivity into and the disappearance of radioactivity from the enzyme.

(a) From the results of your experiments, which are shown in Figures 5-20(b) and (c), respectively, explain the apparent enzyme induction in Figure 5-20(a).

(b) Can you account for these results, if you assume actinomycin D only interferes with transcription?

5-15 Nucleolar RNA from HeLa cells has been isolated and fractionated on sucrose gradients. Different fractions have been examined by electron microscopy under conditions in which the RNA secondary structure is denatured only partially. A typical molecule from the 45S fraction shows numerous base-paired loops along the molecule [Figure 5-21(a)]. Figure 5-21(b) is a representation of ten such 45S molecules aligned for comparison, with the thick lines indicating the base-paired regions. Ten molecules from each of several other fractions also are illustrated; both the 24S and the 20S regions of the gradient had two types of molecules as distinguished by secondary structure. The 36S, 24S(a) and (b), and 20S(a) and (b) molecules all are found in relatively low concentrations.

(a) From these data derive the scheme by which the 45S precursor molecule is cleaved to give 28S and 18S rRNA molecules as well as the other RNA fragments shown.

(b) Is there any evidence for cleavage pathways that do not produce 28S and 18S rRNA?

(c) How might you determine the orientations of these molecules; that is, which end is 5' and which is 3'?

5-16 You have purified a specific mRNA from *Drosophila* cells in tissue culture and you want to localize the gene for that mRNA on polytene chromosomes by *in situ* hybridization [see Essential Concept 2-12(b)]. The mRNA is 1000 nucleotides long and is transcribed from unique DNA. Your preparation of mRNA is labeled with ^{3}H-uridine and has a specific radioactivity of 10^5 disintegrations per minute (dpm)/μg. Assume that the hybridization reaction will go to completion so that each of the 1000 copies of the gene in a polytene chromosome will have an mRNA hybridized to it. (The efficiency of hybridization varies widely with specific experimental conditions; however, in practice it is likely to be much less than the 100% assumed here.) The hybridized chromosomes can be autoradiographed for up to eight weeks and you expect to get 10% efficiency in autoradiography (1 exposed grain/10 disintegrations).

(a) If localization requires that one exposed grain be present reproducibly above one chromomere, will you be able to localize your mRNA under these conditions?

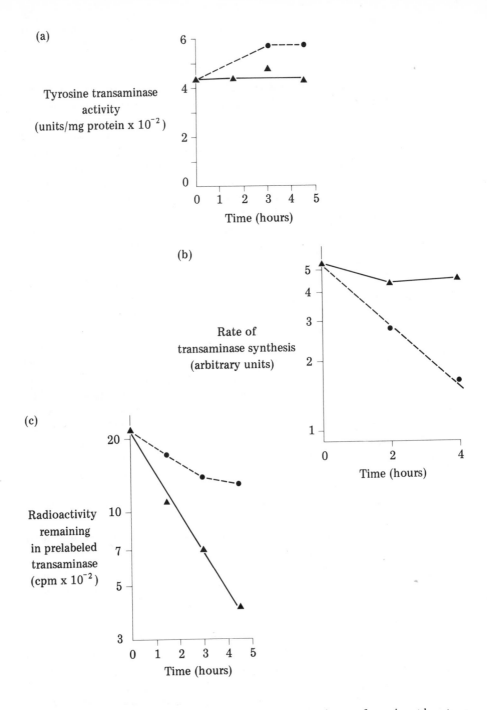

Figure 5-20. Effects of actinomycin D on tyrosine aminotransferase in rat hepatoma cells: (a) enzyme activity; (b) rate of transaminase synthesis; and (c) disappearance of free-labeled transaminase. (-- ●--), actinomycin D added at time 0; (—▲—), no actinomycin D added (Problem 5-14). (Adapted from F. Kinney *et al.*, *Nature New Biol.* **246**, 208, 1973.)

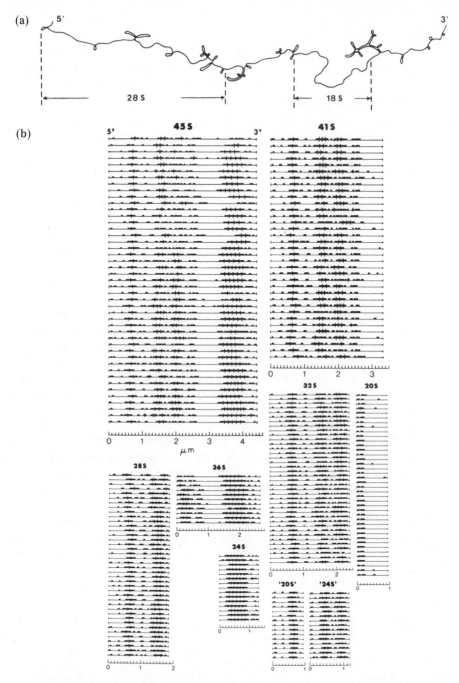

Figure 5-21. Processing of nucleolar RNA from HeLa cells (Problem 5-15). (a) Tracing from an electron micrograph of a typical 45S molecule showing regions of intramolecular base-pairing. (b) Schematic representations showing the arrangement of base-paired regions in molecules from the various gradient fractions of nucleolar RNA. Base-paired regions are indicated by heavy line segments. (Adapted from P. K. Wellauer and I. B. Dawid, *Proc. Natl. Acad. Sci. (U. S.)* **70**, 2827, 1973.)

(b)   What would the specific radioactivity (dpm/μg) of the mRNA have to be in order to localize it on metaphase chromosomes? Assume the same efficiencies of hybridization and autoradiography.

5-17   Members of the picornavirus family, such as poliovirus and encephalomyocarditis virus, contain a single-stranded RNA genome. Initially these genomes were assumed to be polycistronic mRNA's, which were translated to yield several discrete proteins, as are bacteriophage RNA's. However, it has been demonstrated for several picornaviruses that the RNA is actually monocistronic; that is, its translation yields a single huge protein molecule. This protein subsequently is cleaved to form specific polypeptides with distinct functions. The elaborate mechanism that these and other RNA animal viruses employ to produce discrete proteins suggests that animal cells may be incapable of properly translating polycistronic mRNA.

You have been studying a picornavirus that contains single-stranded RNA sufficient to code for about 250,000 daltons of protein. This virus produces five proteins, A, B, C, D, and E, which have the molecular weights shown in Table 5-5. Pulse-chase radioactive labeling experiments have revealed that proteins A, B, and E initially are synthesized in equimolar amounts and that one of these proteins is slowly processed to produce C and D. You suspect that proteins A, B, and E are cleaved from one large precursor, but you have been unable to detect any such precursor in your pulse-chase experiments. To test whether these proteins are synthesized independently, or as part of a common precursor, you carry out the two following experiments.

I.   You add a mixture of radioactive amino acids to infected cells for one hour under conditions where the synthesis of the putative precursor would take only a few minutes. You then separate the radioactively labeled viral proteins on SDS-polyacrylamide gels and determine the amount of radioactivity in each band (Table 5-5).

II.   You repeat Experiment I except that you add pactamycin along with the radioactive amino acids. Pactamycin is an antibiotic that at low concentration blocks initiation of protein synthesis but allows previously initiated translation to continue.

Table 5-5.   Protein synthesis directed by a hypothetical picornavirus (Problem 5-17).

Protein	Molecular weight	Experiment I – pactamycin (cpm/band)	Experiment II + pactamycin (cpm/band)
A	100,000	4600	700
B	89,000	2200	740
C	75,000	880	620
D	60,000	700	500
E	38,000	1800	540

(a)   Which protein is the precursor of proteins C and D?

(b)   Are your experimental results consistent with synthesis of proteins A, B, and E via a single large precursor protein?

(c)   If you conclude that there is a precursor, deduce the order of A, B, and E within it.

5-18   Ovalbumin is produced in large amounts in chick oviduct tissues after stimulation with estrogen, which increases the synthesis of ovalbumin mRNA. Because ovalbumin mRNA (2000 nucleotides long) constitutes such a large fraction of the mRNA in a stimulated cell, it can be purified relatively easily. It is transcribed from unique DNA. Design an experiment to determine whether ovalbumin mRNA is transcribed from unique DNA that is contiguous with a repetitive DNA sequence.

5-19   The RNA polymerases of mouse myeloma (antibody-producing) cells have been separated into four distinct species — RNA polymerases I, II, III_A, and III_B. You have decided to study the possibility that each RNA polymerase has a specialized role in the transcription of genetic information. You purify the four polymerases and find that you can distinguish RNA polymerases I, II, and III (though not III_A from III_B) on the basis of their sensitivity to α-amanitin [Figure 5-22(a)]. You now investigate the sensitivities of various kinds of cellular RNA synthesis to α-amanitin concentration. Your experiments on the sensitivities of endogenous nucleolar and nuclear RNA synthesis are shown in Figure 5-22(b) and those on 5S rRNA and tRNA precursor synthesis are shown in Figure 5-22(c).

(a)   Describe the distribution and quantities of the polymerases in the nucleus and nucleolus.

(b)   Do these RNA polymerases have specialized roles in transcription? If so, describe their roles as thoroughly as your data permit.

5-20   5-Bromodeoxyuridine (BrdU) is incorporated readily into DNA in place of thymidine by all cells. If incorporated into determined but undifferentiated cells, BrdU inhibits the appearance of the differentiated functions that they normally would express (Essential Concept 4-9), but exerts little influence on their proliferation rate or their general viability. For example, erythroid stem cells that have been transformed into leukemic cells by Friend leukemia virus will synthesize hemoglobin if exposed to dimethylsulfoxide (DMSO). However, if BrdU is added to these cells prior to addition of DMSO, the synthesis of hemoglobin is inhibited. BrdU does not seem to affect the rates of total RNA, DNA, or protein synthesis. To study the molecular basis for the BrdU inhibition of hemoglobin synthesis in Friend leukemia cells, you prepare normal mouse-globin mRNA and transcribe it with reverse transcriptase to produce ³H-cDNA. You then prepare total cellular RNA from DMSO-treated and BrdU-DMSO-treated Friend leukemia cells. The results of hybridization of ³H-cDNA with globin mRNA and total cellular RNA's are shown in Figure 5-23(a) and (b), respectively.

(a)   What are the relative concentrations of globin mRNA sequences in the total RNA from DMSO-treated and BrdU-DMSO-treated cells?

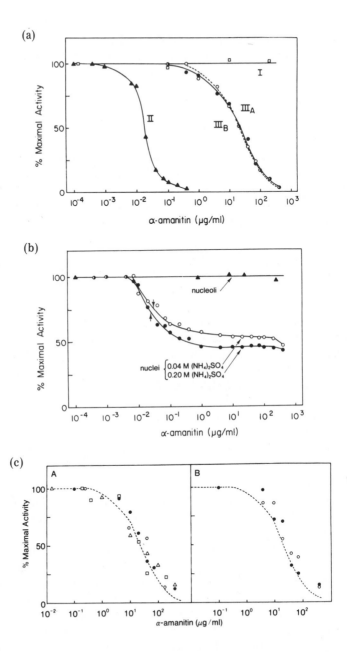

Figure 5-22. Effects of α-amanitin on RNA polymerase activities of mouse myeloma cells (Problem 5-19). (a) Effects of α-amanitin on activity of purified RNA polymerases. (b) Effects of α-amanitin on endogenous RNA polymerase activity in isolated nuclei and nucleoli. (c) Effects of α-amanitin on synthesis of tRNA (Panel A) and 5S RNA (Panel B) in isolated nuclei. (Adapted from R. Weinmann and R. Roeder, *Proc. Natl. Acad. Sci. (U. S.)* **71**, 1790, 1974.)

(a)

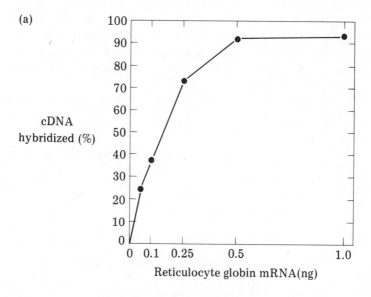

Reticulocyte globin mRNA(ng)

(b)

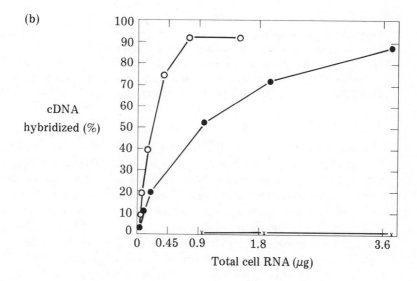

Total cell RNA (μg)

Figure 5-23.   Hybridization of ^{3}H-cDNA with mouse globin mRNA (Problem 5-20). (a) Hybridization with an excess of pure mouse globin mRNA. (b) Hybridization with an excess of total cell RNA from Friend leukemia cells treated with: — ○ —, dimethylsulfoxide; — ● —, BrdU + dimethylsulfoxide; —×—, BrdU alone. The same amount of ^{3}H-cDNA was added to each mixture. (Adapted from H. D. Preisler *et al.*, *Proc. Natl. Acad. Sci. (U. S.)* **70**, 2956, 1973.)

(b)   Did the BrdU treatment lead to a preferential loss of any subset of the globin mRNA sequences normally present in DMSO-treated cultures?

(c)   The size distributions of the globin mRNA's from inhibited and non-inhibited cultures are given in Figure 5-24. Did the BrdU treatment affect the length of the mRNA?

(d)   In view of these results, how can you account for the inhibition of hemoglobin synthesis by BrdU?

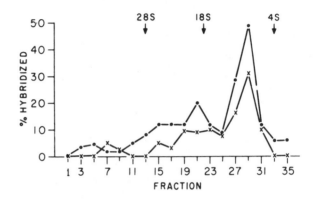

Figure 5-24.   Size distributions of globin mRNA present in total cell RNA from Friend leukemia cells treated with dimethylsulfoxide alone (●) and dimethylsulfoxide + BrdU (×) as determined by sedimentation in sucrose density gradients (Problem 5-20). mRNA was measured by hybridization of fractions from each density gradient to ^{3}H-cDNA as in Figure 5-23. (From H. D. Preisler *et al., Proc. Natl. Acad. Sci. (U. S.)* **70,** 2956, 1973.)

5-21   Consider an organism whose total haploid genome consists of 2×10^9 nucleotide pairs, of which 10^9 nucleotide pairs are unique DNA. In a DNA-driven hybridization experiment, all the mRNA from this organism is shown to be transcribed from unique DNA sequences. When radioactively labeled mRNA from this organism is hybridized at different DNA:RNA ratios to a C_0t of 10^5, the end of the kinetic reaction, the data shown in Figure 5-25 are obtained. Two classes of mRNA are apparent, designated class 1 and class 2 in the figure.

(a)   What is the sequence complexity of the two classes of mRNA's?

(b)   How many different mRNA molecules are there in each class? Assume that each mRNA is 10^3 nucleotides long.

(c)   Calculate the number of molecules per cell for each of the different mRNA's in each class, if there are 4.4×10^{-13} g of mRNA per cell.

5-22   Certain data suggest that each mRNA molecule from *Xenopus laevis* embryos is transcribed partly from unique DNA sequences and partly from repetitive DNA sequences, and that the portion transcribed from repetitive sequences is about 50 to 60 nucleotides long. Given that RNA molecules can be labeled chemically or

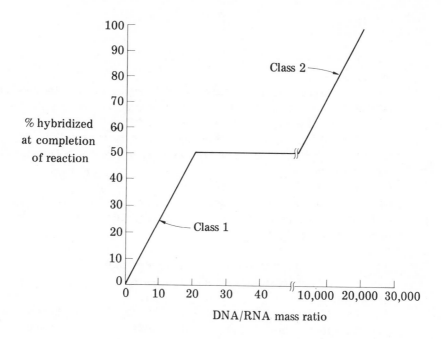

Figure 5-25. Extent of mRNA hybridization as a function of increasing DNA:RNA ratio (Problem 5-21).

enzymatically at the 5' end with ^{32}P, design an experiment that will demonstrate whether the mRNA sequences transcribed from repetitive DNA generally are located at the 5' or 3' ends of mRNA molecules.

5-23 Several investigators have attempted to measure how much of a eucaryotic genome codes for mRNA by hybridization of labeled DNA with an excess of polysomal RNA (presumably mRNA). Generally, such studies have been done under conditions that allow only the more common mRNA's to hybridize. Messengers that are made in only a few cells in the organism, or that are made during a brief period during development, or that are made by all cells but at a very low level would not have hybridized with DNA at the RNA:DNA ratios used.

In a recent study, mRNA was isolated from sea urchin embryos at the gastrula stage (600 cells), and was hybridized in excess with labeled unique DNA. At an $R_0 t$ of 10^3, about 1% of the DNA was hybridized. Assuming asymmetric transcription, the authors concluded that about 2% of the unique DNA is used to code for gastrula message. Is this estimate likely to be an accurate one? Calculate the number of copies of an mRNA that would have to be present in each embryo in order to be half hybridized at $R_0 t = 10^3$. Assume an average of 1200 nucleotides per mRNA, 7.2×10^7 mRNA molecules per embryo, and a K of 5×10^5 (liter nucleotides/mole seconds).

ANSWERS

5-1　(a)　True

(b)　True

(c)　False. Generally, an undetermined cell becomes committed sequentially to more and more specific developmental subprograms. It is only the last such subprogram that produces a specific differentiated cell type.

(d)　False. A stem cell may be determined to give rise to a specific cell type, but stem cells are considered to be undifferentiated since they do not express differentiated functions.

(e)　True

(f)　False. Studies with allophenic mice and *Drosophila* gynandromorphs suggest that most tissues develop from multiple progenitor cells that become committed independently.

(g)　True

(h)　True. (Certain thoracic muscles degenerate in these mutants.)

(i)　False. Allophenic mice are normal in size.

(j)　True

(k)　False. Since liver cells from allophenic mice are of one or the other parental type, the different forms of isocitrate dehydrogenase are not mixed together in the same cytoplasm. Hence, hybrid forms of the enzyme would not be expected even if the enzyme contains multiple subunits.

(l)　True

(m)　False. The determined state of an imaginal disc usually is extremely stable. Changes in this commitment (transdetermination) can occur, but do not in the course of normal development.

(n)　False. The adult environment does not supply the hormone ecdysone, which is necessary to trigger the differentiation of imaginal discs.

(o)　True

(p)　False. There appear to be only a few RNA polymerases in eucaryotic cells and they do not show tissue specificity.

(q)　False. In the case of the 5S rRNA genes of *Xenopus, E. coli* RNA polymerase preferentially transcribed the wrong strand of DNA as well as the normally untranscribed spacer region. In general, the fidelity of eucaryotic transcripts made with *E. coli* RNA polymerase must remain suspect until verified.

(r)　True

(s)　True

(t)　False. Only the very early puffs are activated directly by ecdysone. These early puffs presumably produce proteins that activate the late puffs, thereby triggering a series of developmental programs.

(u)　True

(v)　True

(w)　False. mRNA in procaryotes is very often polycistronic.

(x)　False. Membrane-bound ribosomes synthesize predominantly membrane proteins and proteins for secretion.

5-2   (a)   Determination
      (b)   differentiation
      (c)   allophenic
      (d)   imaginal discs
      (e)   transdetermination
      (f)   homoeotic
      (g)   hnRNA
      (h)   gene amplification
      (i)   Membrane-bound
      (j) α-amanitin
      (k)   Nonhistone chromosomal
      (l)   RNA polymerase I
      (m)   effector

5-3   It is unlikely that histones are the primary regulators of gene expression, because they are not sufficiently heterogeneous to differentially control specific genes (see Essential Concept 2-5). Histones also do not exhibit tissue or organ specificity.

5-4   (a)   Differentiated
      (b)   Differentiated
      (c)   Differentiated
      (d)   Determined
      (e)   Differentiated
      (f)   Determined
      (g)   Determined (Hematopoietic stem cells are committed to produce one of three cell types.)
      (h)   Determined (The transdetermined state is stable.)

5-5   (a)   Since each of the 15 allophenic mice expressed only one of the two alleles, the entire system of blood cells must have been derived from a single progenitor cell. This tissue system apparently is the only one in mammals that has a single progenitor cell rather than multiple, independently determined progenitor cells.
      (b)   The probability that both progenitor cells would have the same allele, Y^a or Y^b, would be 1/2. If the entire system were derived from two progenitor cells, the probability of selecting 15 consecutive mice that each express one or the other allele would be $(1/2)^{15}$ or 1/32,768 (about 3×10^{-5}). If each of the three cell types were derived from two progenitor cells, the probability of obtaining the results in Table 5-3 would be $(1/32,768)^3$. These very low probabilities strongly support the conclusion in Part (a).

5-6   A wild-type head surface does not mean necessarily that the brain underneath is wild-type. Flies that have a normal head surface but drop dead appear to have at least partially mutant brains. The lack of perfect correlation between surface markers and the genotype of the underlying tissue is a source of ambiguity in analyzing gynandromorphs [see Essential Concept 5-3(b)].

5-7 (a) The three bands in the electrophoretic pattern for a mouse that is heterozygous for the two alleles of isocitrate dehydrogenase [Figure 5-17(d)]
indicate that the enzyme is probably a dimer. A hybrid enzyme can be formed
only in cells that contain both alleles. Mixing pure forms of each variant or tissues from homozygous individuals does not produce a hybrid enzyme.

(b) Skeletal muscle is a *syncytium* (a tissue composed of multinucleate cells),
which apparently is formed by the fusion of cells of each parental type. Thus in
skeletal muscle cells a single cytoplasm contains both types of nuclei and a
hybrid enzyme can be formed. No other mouse tissue seems to be formed by cell
fusion (including the heart, which also is syncytial).

5-8 Among a substantial number of allophenic mice, 1/4 were nonmosaic; that is,
they developed from cells of only one of the two genotypes present in equal
numbers at the blastocyst stage. Assuming that blastocyst cells become mixed
randomly before the determination of cells for the embryo, the probability, P, of
determining n cells that all are of the same genotype is given by the expression

$$P = (1/2)^{n-1}$$

From the experimental data, $P = 1/4$, so that

$$(1/2)^{n-1} = 1/4$$

or

$$2^{n-1} = 4$$

and

$$n = 3$$

Thus the data suggest that the embryo most probably arises from only three
progenitor cells in the blastocyst.

5-9 (a) At the rate of 3×10^6 ribosomes per day it would take an egg $10^{12}/3 \times 10^6$
$= 3.3 \times 10^5$ days or approximately 1000 years to synthesize the required number of ribosomes.

(b) An egg increases its rate of ribosome production primarily through gene
amplification. The 1000-fold amplification of rRNA genes increases ribosome
production by a factor of about 1000. In addition, each of these genes is transcribed at close to the maximum rate, whereas a typical somatic cell produces rRNA at
only a fraction of the maximum rate.

(c) Cells can produce specific proteins at a tremendous rate because the transcription product (mRNA) can be reused many times. For example, a single silk gland cell produces 10^5 silk fibroin mRNA's in four days, and each mRNA is translated 10^5 times, on the average, to produce 10^{10} silk fibroin molecules. Thus the rate of silk fibroin production depends not only on the rate of mRNA synthesis, but also on the repeated use of mRNA. By contrast, rRNA is used stoichiometrically; that is, one transcript is used to make one ribosome.

5-10 If 0.070% of the DNA sequences code for 28S rRNA, then considering the complementary DNA strands for these regions, 0.14% of the DNA consists of 28S rRNA genes of molecular weight 3.2×10^6. Thus the diploid genome of *Xenopus laevis* must contain

$$\frac{(3.6 \times 10^{12} \text{ daltons/genome}) (1.4 \times 10^{-3})}{3.2 \times 10^6 \text{ daltons/rRNA gene}}$$

$$= 1600 \text{ rRNA genes/genome}$$

5-11 The absence of G residues from poly-A segments isolated after T1 ribonuclease digestion, and the absence of U and C residues after pancreatic ribonuclease digestion, indicate that the poly-A segment must constitute the 3′ termini of hnRNA and mRNA.

5-12 (a) One gene (10^6 daltons) has an average of 1.08×10^6 daltons of histone and 7×10^5 daltons of nonhistone chromosomal protein associated with it. Thus there are 90 molecules of histone (1.08×10^6 daltons per gene/12×10^3 daltons per histone) and 10 molecules of nonhistone chromosomal protein (7×10^5 daltons per gene/7×10^4 daltons per nonhistone protein) complexed with an average gene.
(b) From (a), there are 100 molecules of chromosomal protein per gene. Thus there are (100 molecules of chromosomal protein/10^6 daltons of DNA) (10^{13} daltons of total DNA) $= 10^9$ molecules of chromosomal protein.
(c) If there is one activator molecule for each of the 2000 structural gene sequences in the battery, then this protein will represent $2000/10^9 = 2 \times 10^{-4}\%$ of the total chromosomal protein population. Isolation of a protein present in such low amounts would be extremely difficult.

5-13 (a) The specific activity of tryptophan pyrrolase after stimulation, as in Experiment I, is a measure of its rate of synthesis. The increase in specific activity after stimulation by hydrocortisone indicates that hydrocortisone increases the enzyme's rate of synthesis. The disappearance of radioactive tryptophan pyrrolase, as measured in Experiment II, is an indication of its rate of degradation. The decreased rate of disappearance in the presence of tryptophan indicates that tryptophan increases the activity of tryptophan pyrrolase by decreasing its rate of degradation. (It is a common observation that the stability of an enzyme is increased in the presence of a cofactor or substrate.)
(b) Tryptophan causes a constant percentage decrease in the rate of tryptophan

pyrrolase degradation, thereby producing a constant percentage increase in the number of enzyme molecules. However, the absolute increase in total activity of tryptophan pyrrolase depends on the number of enzyme molecules present. Since there are more enzyme molecules present after hydrocortisone stimulation, tryptophan causes a larger increase in total activity.

(c)   Actinomycin D would not be expected to affect the results obtained with tryptophan. However, it would affect the hydrocortisone results, if the increase in tryptophan pyrrolase synthesis required mRNA synthesis. As indicated in Essential Concept 6-8, most steroid hormones stimulate specific protein synthesis by stimulating mRNA synthesis. Therefore, actinomycin D would be expected to eliminate the hydrocortisone-induced increase in tryptophan pyrrolase activity.

5-14   (a)   In the absence of actinomycin D, the total enzyme activity remains constant. Thus the rate of synthesis in Figure 5-20(b) must be equal to the rate of degradation in Figure 5-20(c); that is, half the enzyme is replaced every two hours. In the presence of actinomycin D, enzyme degradation is decreased to about a third the previous rate. Thus enzyme synthesis outstrips enzyme degradation until the synthesis rate drops to about 1/3 the control rate (at three hours). Consequently, the total enzyme activity increases to a maximum at about three hours.

(b)   The effect of actinomycin D on the rate of enzyme degradation is difficult to account for, if it is assumed that actinomycin D blocks only transcription. Conceivably, the results could be explained by postulating a labile degradation factor that normally is synthesized continuously. In the presence of actinomycin D this labile factor could not be synthesized and its concentration would decrease rapidly, thereby decreasing the rate of enzyme degradation. There is no evidence for such a factor at present. An alternative possibility is that actinomycin D affects other cellular processes besides transcription.

5-15   (a)   See Figure 5-26.

(b)   The cleavage of 32S to 20S(b) and 24S(b) cuts the RNA in the middle of what could otherwise be a 28S rRNA. This cleavage must represent a nonproductive pathway or an artifactual degradation produced during the isolation process.

(c)   You could digest partially any class of molecules with an exonuclease that attacks only one end of an RNA chain. Examination of the secondary structure of the remaining fragments would show which end has been digested off.

5-16   (a)   At 100% efficiency of hybridization, there will be (10^3 genes) (10^3 nucleotides bound/gene) = 10^6 nucleotides of mRNA bound per chromosome. This value corresponds to (10^6 nucleotides) (330 g/6.0×10^{23} nucleotides) (10^6 μg/g) = $5.5 \times 10^{-10}\mu$g, which in 8 weeks would expose (5.5×10^{-10} μg/chromosome) (10^5 dpm/μg) (8.1×10^4 min) (1 grain/10 disintegrations) = 0.44 grains/chromosome. This value approaches the limit of detectability; examination of 100 chromosome sets should show, on the average, about 44 sets with an exposed grain over a particular chromomere.

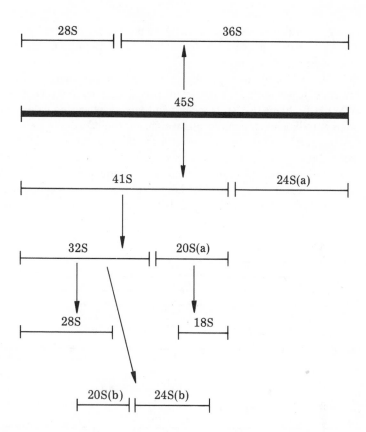

Figure 5-26. Stages in the processing of HeLa ribosomal RNA (Answer 5-15). (Adapted from P. Wellauer and I. B. Dawid, *Proc. Natl. Acad. Sci. (U. S.)* **70,** 2827, 1973.)

(b) Since the corresponding location on a metaphase chromosome contains only one gene copy, the mRNA must have a 1000-fold higher specific radioactivity to reach the level of detectability obtained with polytene chromosomes in Part (a); therefore, the specific radioactivity of the mRNA would have to be 10^8 dpm/μg.

5-17 (a) Since the labeling time in Experiment I is long relative to the synthesis time, all of the proteins will be uniformly labeled, and the amount of radioactivity in each protein band should be directly proportional to its molecular weight, *if* all of the proteins are present in equimolar amounts. Proteins A and E have ratios of 46 and 47 cpm/1000 daltons, respectively, whereas protein B has a ratio of only 25 cpm/1000 daltons. These calculations suggest that some of protein B may have been cleaved to produce proteins C and D. In support of this possibility, if the radioactivity in proteins C and D is corrected for the presumed degraded pieces of B (89,000 − 75,000 = 14,000 daltons for C and 89,000 − 60,000 = 19,000 for

D) and is added to the radioactivity in protein B, the total gives a ratio for uncleaved B of about 48 cpm/1000 daltons. Thus proteins C and D are produced by processing of protein B.

(b)   The possibility of a precursor can be decided by analysis of the data in the presence of pactamycin. Experiment II is analogous to the Dintzis experiment (WWBH Problem 18-18). If the proteins were synthesized independently, their ratios of radioactivity to molecular weight all would be the same. However, if the proteins were synthesized as part of a precursor, their ratios of radioactivity to molecular weight would vary, depending on their location in the precursor. The two possibilities are diagramed schematically in Figure 5-27(a). The actual ratios in Experiment II are 7 cpm/1000 daltons for protein A, 14 cpm/1000 daltons for protein E and about 21 cpm/1,000 daltons for protein B. (Since it is not known

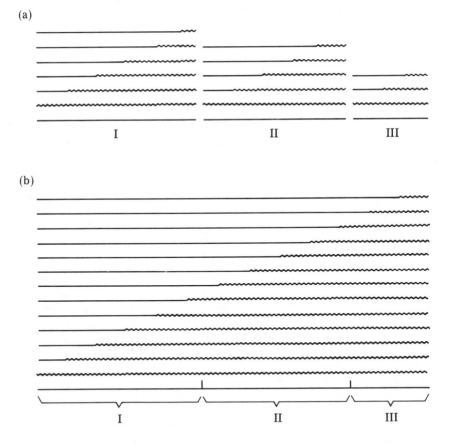

Figure 5-27.   Distribution of radioactivity in the presence of pactamycin in (a) polypeptides that are synthesized independently and (b) polypeptides that are synthesized as part of a larger precursor (Answer 5-17). Straight and wavy lines represent unlabeled and labeled polypeptide sequences, respectively.

which end of protein B is degraded in processing, an exact correction cannot be made.) These results indicate that proteins A, B, and E are cleaved from a common precursor.

(c)   As is indicated in Figure 5-27(b), the ratio of radioactivity to molecular weight should increase from the N- to the C-terminus of the precursor. Thus the order of the proteins in the precursor must be A-E-B.

5-18   Long segments of DNA that contain repetitive sequences can be isolated by shearing chick DNA to fragments of 2000-3000 nucleotide pairs in length, denaturing, and then reassociating to a C_0t value just sufficient to anneal repetitive sequences. Fragments containing repetitive sequences can be recovered from this reassociation mixture by passing it over hydroxyapatite and isolating the fraction that is retained. Given labeled ovalbumin mRNA, a DNA-excess hybridization experiment with the isolated DNA fraction will indicate whether it contains unique sequences complementary to the mRNA.

5-19   (a)   The endogenous RNA polymerase activity in isolated nucleoli is insensitive to concentrations of α-amanitin that completely inhibit RNA polymerases II and III. Thus nucleoli must contain almost exclusively RNA polymerase I. The sensitive fraction of the endogenous RNA polymerase activity in isolated nuclei shows the same sensitivity to α-amanitin as RNA polymerase II. Thus most of the sensitive activity must be due to RNA polymerase II. (The insensitive fraction of endogenous RNA synthesis in nuclei must be due to RNA polymerase I in nucleoli, since the RNA synthesized under these conditions is entirely ribosomal.) Since little if any additional inhibition is seen at the concentration of α-amanitin that completely inhibits RNA polymerases III_A and III_B, their location is undetermined. These results suggest that RNA polymerases I and II each account for about 50% of the total nuclear activity, and that RNA polymerases III_A and III_B account for a very small fraction of the total nuclear activity.

(b)   Since RNA polymerase I is the predominant polymerase activity in the nucleolus, it must be responsible for 18S and 28S rRNA synthesis. Since synthesis of 5S rRNA and precursor tRNA is inhibited at α-amanitin concentrations that inhibit RNA polymerases III_A and III_B, their synthesis must be due to one or the other or both of these polymerases. Since RNA polymerase II accounts for most of the nonnucleolar RNA synthesis, it must be responsible for most hnRNA and mRNA synthesis, although RNA polymerases III_A and III_B also could account for some very small percentage of this synthesis.

5-20   (a)   The 50% saturation points of the hybridization curves in Figure 5-23(b) show that the globin mRNA content of the DMSO-treated cultures is about 3 times greater than that from cultures treated with DMSO and BrdU.

(b)   Since the RNA from inhibited and noninhibited cultures reached the same maximum hybridization value, all of the globin mRNA sequences must be present after BrdU-DMSO-treatment, but at lower than normal concentration.

(c)   Since the distributions of hybridizable RNA obtained from inhibited and noninhibited cultures were very similar, there must have been no appreciable change in the length of the mRNA.

(d)   The effects of BrdU on globin mRNA synthesis are quantitative rather than qualitative since similar transcripts are present in both cultures. Thus BrdU appears to affect only the rate of transcription of the gene. Recently, it has been reported that the rate of dissociation of *lac* repressor from BrdU-substituted *lac* operator DNA in *E. coli* is ten times slower than normal. Conceivably, BrdU could have a similar effect on the dissociation of some regulatory protein in eucaryotic cells.

5-21   (a)   The class 1 mRNA is hybridized completely at a DNA:RNA mass ratio of 40:1. At this point the mass ratio of unique DNA, which contains all of the mRNA sequences, to mRNA, is 20:1. Assuming transcription of only one DNA strand, the ratio of total unique DNA sequences in nucleotide pairs to RNA sequences in class 1 is 10:1; that is, one tenth of the unique DNA sequences are represented in the class-1 mRNA population. Given that the complexity of the unique DNA is 10^9 nucleotide pairs, the complexity of the class-1 mRNA is 10^8 nucleotides. By similar reasoning, the complexity of the class-2 mRNA is 10^5 nucleotides.

(b)   If the average mRNA molecule is 10^3 nucleotides long, there must be $10^8/10^3 = 10^5$ different mRNA molecules in class 1 and $10^5/10^3 = 10^2$ different mRNA molecules in class 2.

(c)   The total number of mRNA molecules per cell is

$$\frac{(4.4 \times 10^{-13} \text{ g/cell})(6.02 \times 10^{23} \text{ nucleotides/mole of nucleotides})}{(330 \text{ g/mole of nucleotides})(10^3 \text{ nucleotides/mRNA molecule})}$$

or

8.0×10^5 mRNA molecules per cell.

For the 50% of the mRNA that represents 10^5 different kinds of mRNA molecules (class 1), there are

$$\frac{(0.5)(8.0 \times 10^5 \text{ mRNA molecules/cell})}{10^5 \text{ kinds of mRNA}}$$

or

4 mRNA molecules of each kind per cell.

For the 50% of the mRNA that represents 10^2 different kinds of mRNA (class 2) a similar calculation shows that there are 4×10^3 mRNA molecules of each kind per cell.

5-22   There are several ways to determine the location of the repetitive portion of the mRNA. One possibility is to isolate intact mRNA that is uniformly labeled with ^{3}H-uridine and label it with ^{32}P at the 5' end. The ^{32}P-labeled mRNA then could be fragmented into small pieces and analyzed in a DNA-driven hybridization reaction. Treatment of the hybrids with ribonuclease should show that the ^{3}H-mRNA reacts with two DNA components. A small fraction should hybridize with the $C_0 t_{\frac{1}{2}}$ of repetitive DNA, whereas the majority should hybridize with the $C_0 t_{\frac{1}{2}}$ of unique DNA. By contrast, the ^{32}P label should hybridize entirely with either the $Cot_{\frac{1}{2}}$ of repetitive DNA or unique DNA. If the portion of mRNA transcribed from repetitive DNA is located on the 5′ end, ^{32}P label will hybridize entirely with repetitive DNA. If the repetitive portion is located internally or on the 3′ end, ^{32}P label will hybridize entirely with unique DNA.

5-23   The average complexity of any single mRNA species will be equal to its chain length; that is, $X = 1200$ nucleotides. The $R_0 t_{\frac{1}{2}}$ for a pure preparation of such an RNA can be calculated using Equation A5-11 as follows:

$$R_0 t_{1/2 \text{(pure)}} = \frac{(X)(\ln 2)}{2K}$$

$$= \frac{(1200 \text{ nucleotides})(0.69)}{2 \times 5 \times 10^5 \text{ liter nucleotides/mole sec}}$$

$$= 8.3 \times 10^{-4} \text{ mole sec/liter}$$

If this same RNA is present as some fraction, f, of the total RNA population, its $R_0 t_{\frac{1}{2}}$, in terms of total RNA concentration, will be higher by the factor $1/f$, that is,

$$R_0 t_{1/2 \text{(mixture)}} = \frac{R_0 t_{1/2 \text{(pure)}}}{f}$$

or

$$f = \frac{R_0 t_{1/2 \text{(pure)}}}{R_0 t_{1/2 \text{(mixture)}}}$$

For an RNA species to be half hybridized at $R_0 t = 10^3$ mole sec/liter, f must be $8.3 \times 10^{-4}/10^3 = 8.3 \times 10^{-7}$. For the organism in the problem, this fraction represents $(8.3 \times 10^{-7})(7.2 \times 10^7$ mRNA's per embryo$) = 60$ copies per embryo, or 0.1 copies per cell. Therefore hybridization to $R_0 t = 10^3$ should detect any mRNA that is common to all the cells of the gastrula or that is made in at least 60 copies by a subpopulation of gastrula cells. Hybridization to this value of $R_0 t$ will not detect an mRNA that is made in less than 60 copies by a subpopulation of gastrula cells. Thus the conclusion that 2% of the unique DNA is used to code for gastrula mRNA may be an underestimate of the true value.

6 Hormone Action

In multicellular organisms, the cells that first receive a stimulus and the cells that respond physiologically are often in different tissues. Signals are transmitted between these cells by electrical impulses and chemical messengers. These two mechanisms are used separately or in series for all intercellular communication. This chapter considers an important class of chemical messengers, the hormones, and how they act in higher animals.

ESSENTIAL CONCEPTS

6-1   Hormones carry signals from one set of cells to another set via the circulatory system

(a)   Hormones are released into the circulatory system by one set of cells (generally *endocrine* cells) in order to influence the physiological behavior of other cells (*target* cells). Hormones affect target cells either by altering the properties of existing proteins or by initiating the synthesis of new ones. Thus hormones tune the metabolism of individual cells for the optimal function of the organism.

(b)   The principal mammalian hormones are listed in Table 6-1 with their structures and primary physiological effects. The hierarchy of their regulation is shown schematically in Figure 6-1.

    (1)   As indicated in Figure 6-1, the hypothalamus exerts primary control by secreting releasing and inhibiting factors in response to neuronal and chemical stimuli. These factors act primarily on the anterior pituitary (adenohypophesis) to control the secretion of hormones that exert a variety of effects on specific tissues. The secretion of the hypothalamic regulatory factors and the anterior pituitary hormones is subject to feedback inhibition.

    (2)   Hormone secretion by the posterior pituitary (neurohypophysis) and the adrenal medulla is controlled more directly by the central nervous system. The posterior pituitary hormones actually are synthesized in the hypothalamus. They migrate down the hypothalamic nerves that innervate the hypothalamus, and are secreted from the nerve endings in response to appropriate neuronal stimulation. The adrenal medulla is stimulated directly by the central nervous system to secrete epinephrine and norepinephrine, the hormones that elicit the well known "fight or flight" response.

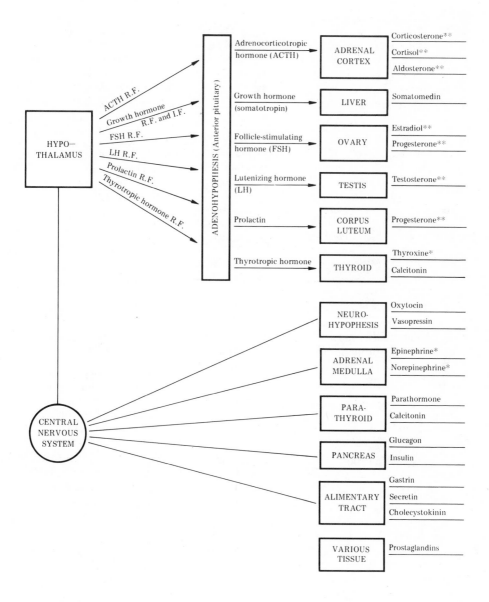

Figure 6-1. The hierarchy of hormone regulation in mammals. The initials R.F. and I.F. stand for releasing factors and inhibiting factors, respectively. Arrows indicate hormone action. Hormones with no asterisk are polypeptides, those with one asterisk are amino acid derivatives, and those with two asterisks are steroids.

Table 6-1. Structures and physiological effects of principal mammalian hormones.

Hormone	Structure	Primary physiological effects
Adrenocorticotropic hormone	polypeptide, 39 residues	Lipid release from adipose tissue; stimulation of adrenal cortex to produce corticosterone, cortisol, and aldosterone
Corticosterone		Metabolism of proteins, carbohydrates, and lipids; maintenance of circulatory and vascular homeostasis; inflammation, immunity and resistance to infection; hypersensitivity
Cortisol		Same as corticosterone
Aldosterone		Maintenance of electrolyte and water balance
Growth hormone (somatotropin)	polypeptide, 21,500 MW	Growth of bone and muscle; anabolic effect on Ca^{2+}, phosphate, and nitrogen metabolism; metabolism of carbohydrate and lipid; elevation of muscle and cardiac glycogen; stimulation of liver to produce somatomedin
Somatomedin	polypeptide, 8,000 MW	Mediates some if not all growth hormone effects
Follicle-stimulating hormone	glycoprotein, 34,000 MW α and β subunits, 16% carbohydrate	Development of follicles and ovulation in ovary; development of seminiferous tubules and spermatogenesis in testis; stimulation of ovary to produce estradiol
Estradiol		Development of secondary female sex characteristics; maturation and normal cyclic function of accessory sex organs; development of duct system in mammary glands

Table 6-1. (Continued)

Hormone	Structure	Primary physiological effects
Lutenizing hormone	glycoprotein, 28,500 MW α and β subunits, 15.5% carbohydrate	Lutenizing in ovary; development of interstitial tissue in testis; stimulation of ovary to produce progesterone and estradiol; stimulation of testis to produce testosterone
Progesterone		Preparation of uterus for ovum implantation; maintenance of pregnancy; development of alveolar system in mammary glands
Testosterone		Development of secondary male sex characteristics; maturation and normal function of accessory sex organs
Prolactin	polypeptide, 23,500 MW (bovine)	Proliferation and initiation of milk secretion in mammary gland; proliferation of corpora lutea; stimulates corpus luteum to produce progesterone
Thyrotropic hormone	glycoprotein, 28,300 MW (bovine) α and β subunits, 15% carbohydrate	Lipid release from adipose tissue; stimulation of thyroid to produce thyroxin and calcitonin
Thyroxine		Metabolic rate and O_2 consumption
Calcitonin	polypeptide, 32 residues	Metabolism of Ca^{2+} and phosphate
Oxytocin	Cys-Tyr-Ilu-Gln-Asn-Cys-Pro-Leu-Gly (S-S)	Contraction of smooth muscle, particularly uterine; parturition; ejection of milk by mammary gland
Vasopressin	Cys-Tyr-Phe-Gln-Asn-Cys-Pro-Arg-Gly (S-S)	Blood pressure of arterioles; water resorption in kidney tubules

Table 6-1. (Continued)

Hormone	Structure	Primary physiological effects
Epinephrine		Pulse rate and blood pressure; contraction of most smooth muscle; glycogenolysis in liver and muscle; lipid release from adipose tissue
Norepinephrine		Contraction of arterioles, increased peripheral resistance; lipid release from adipose tissue
Parathormone	polypeptide, 9500 MW	Metabolism of Ca^{2+} and phosphate in skeleton, kidney, and gastro-intestinal tract
Glucagon	polypeptide, 29 residues	Glycogenolysis in liver; lipid release from adipose tissue
Insulin	polypeptide, A chain, 21 residues; B chain, 30 residues	General utilization of carbohydrate and stimulation of protein synthesis; lipogenesis in adipose tissue
Gastrin	polypeptide, 17 residues	Secretion of gastric acid and pepsin in stomach
Secretin	polypeptide, 27 residues	Secretion of pancreatic digestive enzymes
Cholecystokinin	polypeptide, 33 residues	Secretion of pancreatic digestive enzymes; contraction and emptying of gall bladder
Prostaglandins	See Figure 6-4	Alteration of blood pressure; contraction of smooth muscle; lipid metabolism in adipose tissue; blood platelet aggregation; may modify responses to synaptic transmitters in brain, mediate inflammatory responses, inhibit gastric secretion by gastro-intestinal tract

(3)   A number of other tissues, including pancreas, parathyroid, and gastrointestinal tract, secrete a few kinds of specific hormones in response to appropriate primary neuronal and chemical stimuli.

6-2   Polypeptides, amino acid derivatives, steroids, and prostaglandins are the general chemical classes of hormones

(a)   Polypeptide and amino acid-derived hormones can be grouped together on the basis of their derivation from amino acids and their similar modes of action (Essential Concept 6-7). Polypeptide hormones are produced by the cells' normal protein synthesizing machinery. The *catecholamines,* norepinephrine and epinephrine, are synthesized from Tyr as shown in Figure 6-2. The other amino acid-derived hormone, thyroxine, is synthesized by iodination and condensation of Tyr residues in a 650,000-dalton protein, thyroglobulin, and subsequently is released by proteolysis. Condensation of two diiodotyrosine residues yields a thyroxine residue and a Ser or Ala residue (it is not known which).

A number of hormones, including thyroxine, insulin, parathormone, glucagon, adrenocorticotropic hormone, and growth hormone, are synthesized as larger precursor molecules (prohormones) that are cleaved prior to secretion. The biological rationale for this mode of synthesis is not clear. Perhaps thyroxine can be synthesized or stored effectively only as part of a large protein. For the polypeptide hormones, precursor synthesis and cleavage allow rapid production of active hormone from prohormone. Moreover, some hormones produced by cleavage, such as insulin, have thermodynamically unstable three-dimensional structures that may be important in facilitating rapid degradation (see Essential Concept 6-3). In addition, since most prohormones have partial biological activity, intracellular cleavage may represent an evolutionary solution to making a more active molecule.

(b)   The steroid hormones are derived from cholesterol as shown in Figure 6-3. Only the screened steroids are secreted in significant amounts. In different tissues these biosynthetic pathways are regulated so that only hormones characteristic of the tissue are secreted (see Figure 6-1). Secretion of steroid hormones generally is stimulated by polypeptide hormones. At least one effect of such stimulation appears to be an increase in the rate of cholesterol to pregnenolone conversion.

(c)   The common prostaglandins are synthesized from two fatty acids that are required in the diet, as shown in Figure 6-4. Unlike the other classes of hormones, prostaglandins are secreted not only from endocrine cells, but also from the cells of many other tissues.

6-3   Hormones are secreted and inactivated rapidly

Organisms require efficient intercellular communication to respond quickly to changes in internal or external environment. Hormonal messages must be sent

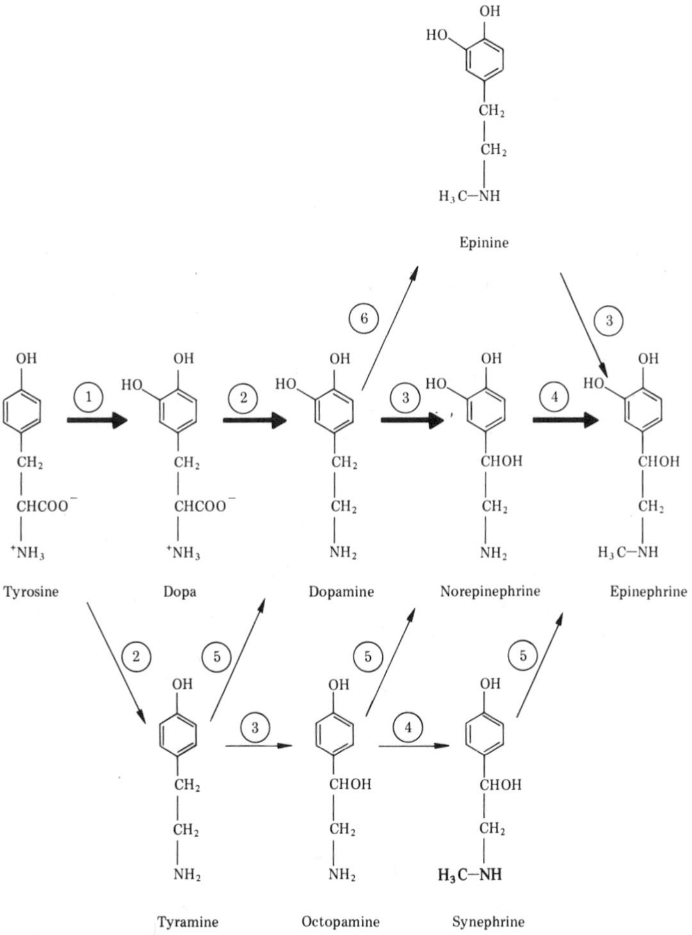

Enzymes

1. Tyrosine hydroxylase
2. Aromatic acid decarboxylase
3. Dopamine-β-hydroxylase
4. Phenylethanolamine-N-methyltransferase
5. Catecholamine-forming enzyme
6. Nonspecific methyl transferase

Figure 6-2. Biosynthesis of catecholamines. Thick arrows indicate the major pathway.

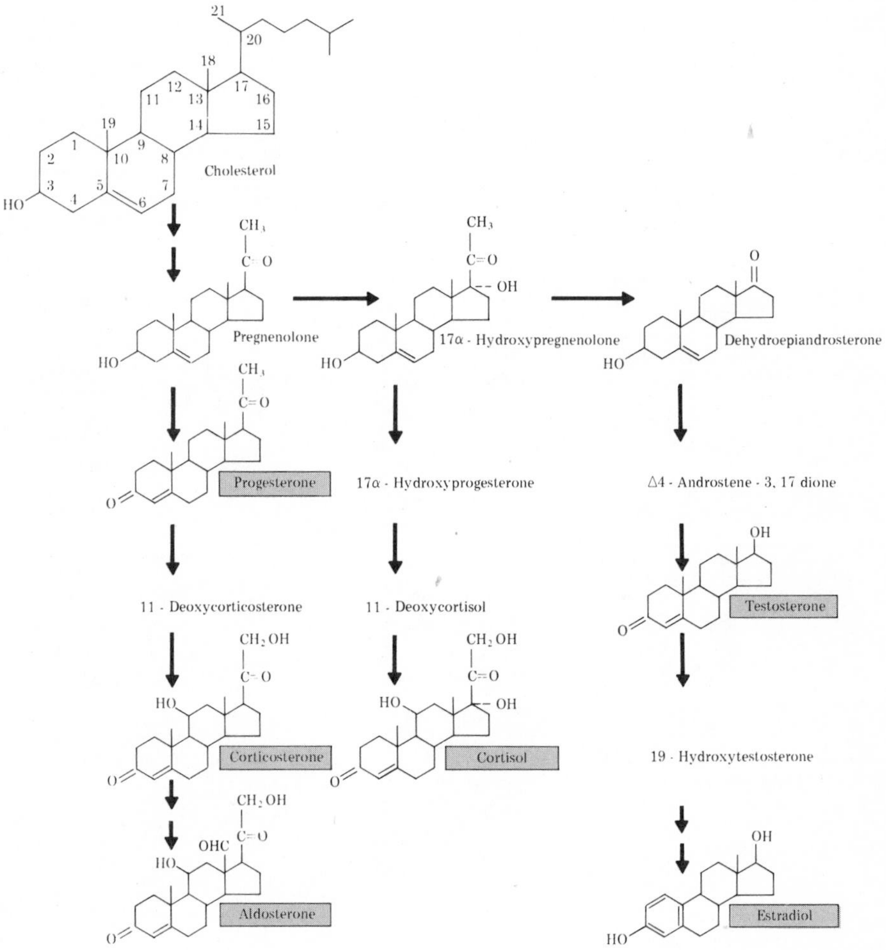

Figure 6-3. Composite outline of steroid biosynthesis. Thick arrows indicate primary pathways. Screened steroids are those that are secreted in significant amounts from endocrine tissues.

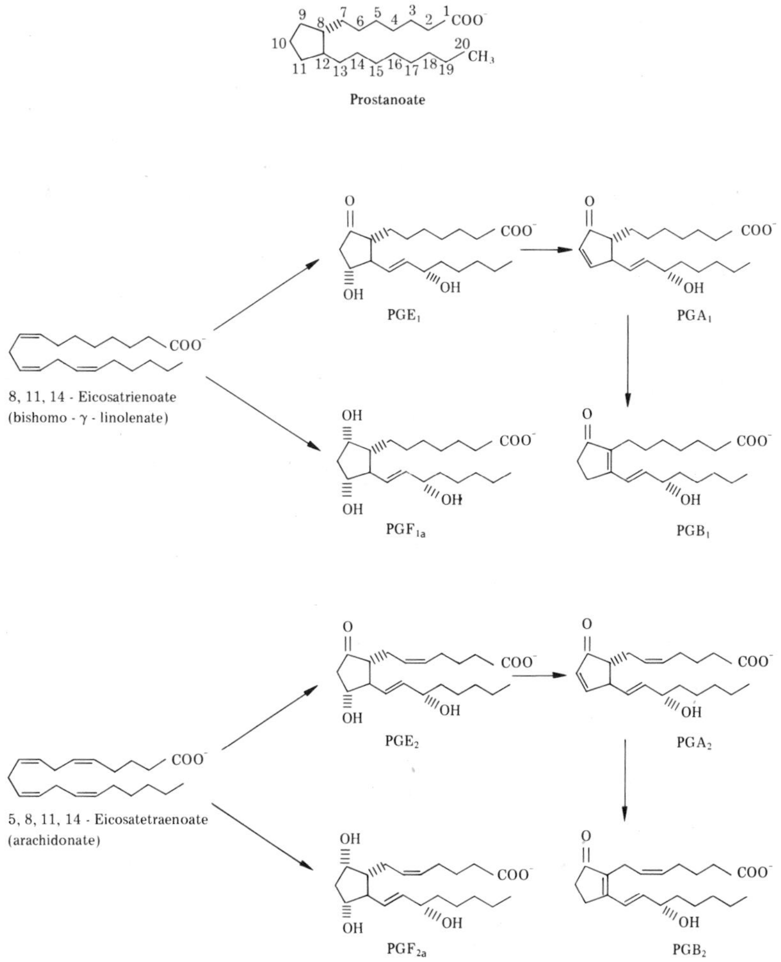

Figure 6-4. Synthesis of prostaglandins. The essential fatty-acid precursors of the prostaglandins (PG) are shown on the left. Bonds directed below the plane of the page are shown as (///////). (Adapted from J. W. Hinman, *Ann. Rev. Biochem.* **41,** 161, 1972.)

with as little lag time as possible, and communication pathways must be kept free of old messages. The first of these requirements is fulfilled by the hormone-producing cells, which are poised to secrete hormones almost immediately in response to stimulation. Polypeptide and amino acid-derived hormones, such as thyroxine and epinephrine, are presynthesized and stored in secretory granules. Steroids and prostaglandins can be synthesized from precursors in a few steps. The second requirement is met by inactivating mechanisms that quickly clear hormones from the circulatory system. All hormones have relatively short physiological half-lives, ranging from two minutes for thyrotropic hormone releasing factor to an hour for cortisol. In general, polypeptide hormones are inactivated by proteolysis, whereas steroids, prostaglandins, and amino acid-derived hormones are inactivated by modification and breakdown.

6-4   Hormones produce quick responses that may be of long or short duration

Hormones that adjust rates of metabolic reactions, such as prostaglandins and most polypeptide hormones, generally produce a quick response of short duration. The magnitude of the response at any given instant is related closely to the concentration of circulating hormone. In contrast, hormones that affect rates of specific protein synthesis, such as most steroids, necessarily produce a somewhat slower response. However, the response is generally longer in duration, depending upon the stability of the newly synthesized protein. Since the prolonged absence of the second class of hormones causes some cells and tissues not to develop properly and others to become incapable of responding to hormones of the first class, these hormones sometimes are referred to as *developmental*.

6-5   Target cells contain specific hormone receptors

(a)   Although circulating hormones are exposed to a wide variety of tissues, each interacts only with specific target cells. These cells contain *receptor* molecules that specifically recognize the hormone. Hormone receptors bind specific hormones and directly or indirectly trigger a metabolic effect. Each target cell is programmed during development to synthesize particular hormone receptors that mediate a characteristic response when occupied. For example, liver cells, in the process of differentiation, synthesize a receptor that allows epinephrine to specifically alter their glycogen metabolism (Table 6-1). Thus the short-term metabolic or developmental effects elicited by hormones are predetermined during development by the synthesis of different receptors in different cell types.

(b)   The receptors for all polypeptide hormones, amino acid-derived hormones, and prostaglandins are located in the surface membranes of target cells. The insulin receptor in liver and fat-cell membranes is one of the best characterized. About 10,000 insulin receptors are distributed over the cell surface at a density of about 10 per μm^2. The receptor is a glycoprotein with a molecular weight of approximately 300,000 and an axial ratio (length:diameter) of 9. It is oriented asymmetrically in the membrane so that the insulin binding site faces outward. The dissociation constant for insulin binding is $5 \times 10^{-11}M$. The rate constant for association ($8.5 \times 10^6 M^{-1}\ sec^{-1}$) approaches the value expected for a diffusion-controlled reaction. The dissociation rate constant is low (4.2×10^{-4} sec^{-1}) so that the complex has a half-life of about 27 minutes. Neither the hormone nor the receptor is inactivated by binding.

(c)   The receptors for steroid hormones are located in the cytoplasm of target cells. Since steroids are nonpolar molecules, they can pass freely through the cell membrane. The progesterone receptor has been isolated from chick oviduct and characterized. There are on the order of 10^5 receptor molecules per cell. The receptor is an asymmetric protein with an axial ratio of 16 and two subunits, which have molecular weights of 110,000 and 117,000. The dissociation constant for progesterone binding is about $8 \times 10^{-10}M$. Neither progesterone nor its receptor is inactivated by binding.

6-6   Many hormonal effects are mediated through cyclic AMP

(a) $3', 5'$-cyclic AMP (cAMP) is synthesized from ATP by *adenyl cyclase* and degraded to $5'$AMP by a specific phosphodiesterase:

$$ATP \xrightarrow[\substack{\text{adenyl} \\ \text{cyclase}}]{Mg^{2+}} PP_i + cAMP$$

$$cAMP \xrightarrow[\substack{\text{phospho-} \\ \text{diesterase}}]{Mg^{2+}} 5'\ AMP$$

The $\Delta G_0'$ for synthesis of cAMP from ATP is about $+2$ kcal/mole, but the reaction is driven by subsequent hydrolysis of pyrophosphate. The $\Delta G_0'$ for cAMP hydrolysis is about -9 kcal/mole. However, cAMP is stable in the absence of phosphodiesterase. The intracellular concentration of cAMP reflects the relative activities of adenyl cyclase and phosphodiesterase. Normally the concentration of cAMP is regulated primarily by alteration of adenyl cyclase activity, but the concentration also may be manipulated artificially by inhibiting phosphodiesterase with methylxanthines, such as theophylline and caffeine.

(b)   cAMP regulates the rates of several cellular processes. Some of its known effects are listed in Table 6-2. cAMP is found in all animal species investigated as well as in bacteria and a number of other unicellular organisms. cAMP functions only as an allosteric effector, either to increase or decrease a reaction rate, but never as a substrate or product except in its own synthesis or degradation. The concentration of cAMP is controlled primarily, if not solely, by extracellular stimuli. In multicellular organisms, one of the most important of these stimuli is hormones.

Table 6-2. Some known effects of cAMP. (Adapted from E. W. Sutherland, *Science* **177,** 401, 1972.)

Enzyme or process affected	Tissue or organism	Change in activity or rate
Protein kinase[a]	Several	Increased
Phosphorylase	Several	Increased
Glycogen synthetase	Several	Decreased
Lipolysis	Adipose	Increased
Clearing-factor lipase	Adipose	Decreased
Amino acid uptake	Adipose	Decreased
Amino acid uptake	Liver and uterus	Increased
Synthesis of several enzymes	Liver	Increased
Net protein synthesis	Liver	Decreased
Gluconeogenesis	Liver	Increased
Ketogenesis	Liver	Increased
Steroidogenesis	Several	Increased
Water permeability	Epithelial	Increased
Ion permeability	Epithelial	Increased
Calcium resorption	Bone	Increased
Renin production	Kidney	Increased
Discharge frequency	Cerebellar Purkinje	Decreased
Membrane potential	Smooth muscle	Increased
Tension	Smooth muscle	Decreased
Contractility	Cardiac muscle	Increased
HCl secretion	Gastric mucosa	Increased
Fluid secretion	Insect salivary glands	Increased
Amylase release	Parotid gland	Increased
Insulin release	Pancreas	Increased
Thyroid hormone release	Thyroid	Increased
Calcitonin release	Thyroid	Increased
Release of other hormones	Adenohypophysis	Increased
Histamine release	Mast cells	Decreased
Melanin granule dispersion	Melanocytes	Increased
Aggregation	Platelets	Decreased
Aggregation	Cellular slime molds	Increased
Messenger RNA synthesis	Bacteria	Increased
Synthesis of several enzymes	Bacteria	Increased
Proliferation	Thymocytes	Increased
Cell growth	Tumor cells	Decreased

[a]Stimulation of protein kinase is known to mediate the effects of cyclic AMP on several systems and may be involved in many or even most of the effects of cyclic AMP.

6-7 Polypeptide hormones and prostaglandins alter the intracellular concentration of cAMP

(a)   Adenyl cyclase in mammalian cells apparently consists of regulatory and catalytic components that are responsible for specific hormone recognition and cAMP synthesis, respectively. These complexes are localized in the plasma membrane where they can interact with a variety of hormones in the first stage of hormonal regulation of target-cell metabolism. The specific hormone receptor, the regulatory component of the complex, is characteristic of the target cell. Polypeptide and amino acid-derived hormones and the prostaglandins apparently bind to the regulatory component and alter the activity of the associated adenyl cyclase, thereby altering the intracellular concentration of cAMP. The majority of hormones increase the activity of adenyl cyclase, whereas a few, such as insulin and some prostaglandins, decrease its activity.

(b)   cAMP has been called the "second messenger," since it transmits the primary hormonal message from the cell surface to the cell interior. The second messenger concept is illustrated in Figure 6-5. A large amplification of the

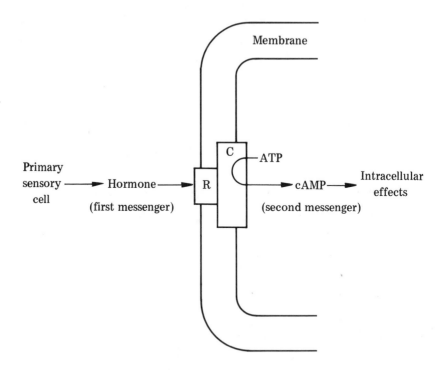

Figure 6-5.   Schematic representation of the second messenger concept. R and C represent the regulatory and catalytic subunits of the adenyl cyclase system. R is target-cell specific, whereas C may be common to all cells.

hormonal signal is achieved by using cAMP as a second messenger. The circulating concentration of many hormones is on the order of $10^{-10}M$, whereas the concentration of cAMP in a stimulated target cell is about $10^{-6}M$. Thus the synthesis of cAMP by activated adenyl cyclase leads to a 10^4-fold amplification of the hormonal signal.

The following experimental criteria have been met in establishing cAMP as a second messenger for a number of hormones.

(1) Adenyl cyclase of broken target-cell preparations is stimulated or inhibited only by hormones that affect the target cell.

(2) The change in cAMP concentration in target cells parallels their biological response to the hormone, both in kinetics and dependence on hormonal concentration.

(3) Inhibitors of the phosphodiesterase act synergistically with hormones that stimulate adenyl cyclase.

(4) The biological effects of the hormone are mimicked by addition of cAMP to the target cell. Since cells are relatively impermeable to cAMP, more hydrophobic derivatives such as dibutyryl-cAMP usually are used for this test.

(c)   The metabolic response of a target cell to a change in cAMP concentration depends upon the cAMP-sensitive regulatory proteins that the cell contains. One of the best understood processes involving cAMP is the breakdown of liver glycogen (WWBH Chapter 9) in response to epinephrine or glucagon (Figure 6-6). The hormonally induced increase in cAMP concentration activates a protein kinase by binding to its regulatory subunit, which activates its catalytic subunit by dissociating from it. The free catalytic subunit phosphorylates enzymes involved in glycogen metabolism. Phosphorylation inactivates glycogen synthetase, a key enzyme in glycogen synthesis, and activates glycogen phosphorylase, a key enzyme in glycogen breakdown. Thus an increase in cAMP concentration promotes glycogen breakdown in the liver by activating glycogen phosphorylase and inactivating glycogen synthetase.

Other responses involving cAMP are less well understood. A current hypothesis proposes that most cAMP effects are mediated by protein kinases. According to this hypothesis, different kinds of target cells respond differently because they contain different substrates for protein kinases. Kinases have been found to phosphorylate histones and ribosomal proteins as well as specific enzymes. Thus second-messenger effects may be exerted at the level of transcription, by selectively removing phosphorylated histones, and at the level of translation, by altering the rate or specificity of protein synthesis, as well as at the level of enzyme activity.

(d)   A second cyclic nucleotide, 3', 5'-cyclic GMP (cGMP), also may mediate cellular responses to external stimuli. Some cellular processes appear to respond oppositely to cAMP and cGMP. As a result, the two cyclic nucleotides have been likened to the Yin and Yang of Chinese philosophy, as opposing effectors of metabolism. The intracellular concentrations of cAMP and cGMP have been found to vary inversely to one another in some cellular responses. According to the Yin/Yang hypothesis, the dual involvement of cAMP and cGMP is restricted to cellular responses that require simultaneous control of two opposing processes, such as glycogen synthesis and breakdown, or cell proliferation and contact inhibition.

6-8 Steroid hormones increase the rates of synthesis of specific proteins

(a)   Unlike the polypeptide hormones, steroids act directly on intracellular components. Steroid hormones stimulate specific protein synthesis in target cells by increasing the rates of specific mRNA production. The sequence of events in this stimulation is diagramed in Figure 6-7. Steroids bind to cytoplasmic receptors in

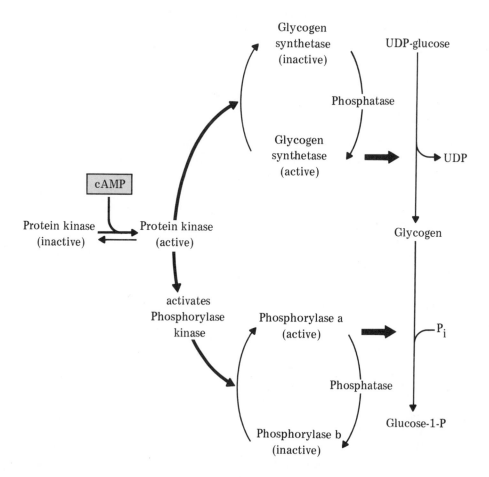

Figure 6-6.   Control of glycogen metabolism by cAMP.

their target cells to form a complex with a sedimentation coefficient of about 8S at low ionic strength and 4S at high ionic strength. This reversible conversion from 4S to 8S presumably is due to aggregation and serves as a useful identifying characteristic of steroid receptors. The 4S complex undergoes a time- and temperature-dependent conversion to a 5S complex that can enter the nucleus and bind to specific sites on chromatin. The 4S → 5S conversion probably results from binding of an additional protein subunit to the 4S hormone-receptor complex. Target-cell chromatin shows seven to eight times greater capacity to bind the 5S complex than does nontarget-cell chromatin, suggesting that target cells may possess a specific chromatin receptor for the 5S complex as well as a cytoplasmic hormone receptor.

Studies on a 5S progesterone-receptor complex suggest a mechanism for interaction of 5S complexes with chromatin. The progesterone receptor can be separated into two components that each bind progesterone with high affinity. One component binds to DNA and the other to a nonhistone chromosomal protein, implying that hormone-receptor complexes recognize and act upon two separate chromatin components.

(b)   The binding of 5S complex to chromatin somehow promotes synthesis of nuclear RNA with subsequent accumulation and translation of specific mRNA in the cytoplasm. For example, following estradiol stimulation, chick oviduct shows an increase from 0 to more than 10^5 ovalbumin-specific mRNA's per cell with a corresponding increase in synthesis of ovalbumin (the principal protein of egg white). The molecular details of how binding leads to production of specific mRNA are not known. Superficially, the process is analogous to positively controlled induction of bacterial enzymes (Essential Concept 1-6) where inducer binding at a specific DNA site promotes transcription of the adjacent gene.

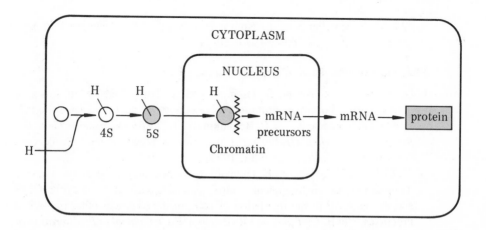

Figure 6-7.   Sequence of events in a steroid hormone response. H stands for hormone.

6-9 Additional concepts and techniques are presented in the Problems section

(a) Models for membrane-bound adenyl cyclase. Problem 6-7.

(b) Mechanisms for opposing effects of cAMP and cGMP. Problem 6-10.

(c) Hormone binding to receptors. Problems 6-14 and 6-16.

(d) Hormone synthesis and degradation. Problem 6-15.

REFERENCES

Where to begin

I. Pastan, "Cyclic AMP," *Scientific American* (August, 1972)

G. S. Stent, "Cellular communication," *Scientific American* (September, 1972)

General

B. Clarkson and R. Baserga (Ed.), *Control of Proliferation in Animal Cells* (Cold Spring Harbor Laboratory, Cold Spring Harbor, N.Y., 1974)

L. Stryer, *Biochemistry* Chapter 34 (Freeman, San Francisco, 1975)

Hormone structure, biosynthesis, and physiology

J. W. Hinman, "Prostaglandins," *Ann. Rev. Biochem.* **41,** 161 (1972)

H. S. Tager and D. F. Steiner, "Peptide hormones," *Ann. Rev. Biochem.* **43,** 509 (1974)

A. White, P. Handler, and E. Smith, *Principles of Biochemistry* Part Five (McGraw-Hill, New York, 1973, 5th ed.)

Molecular mechanisms of hormone action

J. D. Baxter, G. Rousseau, S. Higgins, and G. Tompkins, "Mechanism of glucocorticoid hormone action and of regulation of gene expression in cultured mammalian cells," in P. Benjamin (Ed.), *The Biochemistry of Gene Expression in Higher Organisms* (Australia & New Zealand Book Co., Artarmon, M. S. W., Australia, 1973)

N. D. Goldberg, M. Haddox, E. Dunham, C. Lopez, and J. Hadden, "The Yin Yang hypothesis of biological control: opposing influences of cyclic GMP and cyclic AMP in the regulation of cell proliferation and other biological processes," in B. Clarkson and R. Baserga (Ed.), *Control of Proliferation in Animal Cells* (Cold Spring Harbor Laboratory, Cold Spring Harbor, N. Y., 1974)

E. V. Jensen and E. R. DeSombre, "Estrogen-receptor interaction," *Science* **182,** 126 (1973)

D. McMahon, "Chemical messengers in development: A hypothesis," *Science* **185,** 1012 (1974)

B. W. O'Malley and A. R. Means, "Female steroid hormones and target cell nuclei," *Science* **183,** 610 (1974)

H. L. Segal, "Enzymatic interconversion of active and inactive forms of enzymes," Science **180,** 25 (1973)

E. W. Sutherland, "Studies on the mechanisms of hormone action," *Science* **177,** 401 (1972)

"Membranes and mechanisms of hormone action," (A series of four symposium papers) *Federation Proceedings* **32,** 1833 (1973)

PROBLEMS

6-1 Answer the following with true or false. If false explain why.

(a) Hormones include all chemical messages that are passed from one cell to another.

(b) Hormones tune the metabolism of individual cells for the proper overall functioning of the organism.

(c) The anterior pituitary (adenohypophysis), posterior pituitary (neurohypophysis), adrenal medulla, and parathyroids are controlled by the central nervous system.

(d) All hormones are presynthesized and stored in secreting granules.

(e) All hormones have relatively short physiological half-lives.

(f) The magnitude of cellular response to prostaglandins and most polypeptides is related closely to the concentration of circulating hormone at any given instant.

(g) All target cells carry hormone receptors at their surfaces.

(h) The insulin receptor is oriented asymmetrically in the membrane so that the insulin binding site faces inward.

(i) A hormone is inactivated by binding to its receptor; the receptor, upon release of one hormone molecule, can bind another.

(j) All the amino acid-derived hormones are derived from Tyr.

(k) Thyroxine is produced by condensation of two diiodotyrosine residues.

(l) All steroid hormones are elaborated in increased amounts in response to stimulation by polypeptide hormones.

(m) cAMP, which has a high negative $\Delta G_0'$ of hydrolysis, sometimes is used to drive reactions that are unfavorable energetically.

(n) Intracellular cAMP concentrations change primarily, if not solely, in response to external stimuli.

(o) Phosphodiesterase systems in mammalian cells apparently include separate subunits for hormone recognition and cAMP degradation.

(p) A large amplification of hormonal signal is achieved by using cAMP as a second messenger.

(q) Theophylline would be expected to act synergistically with insulin.

(r) The biological effects of cortisol can be mimicked by treatment of target cells with dibutyryl cAMP.

(s) All hormones that affect a specific cell by increasing cAMP concentration will have the same effect on that cell.

(t) Steroid hormones increase the rate of synthesis of specific proteins.

(u) The progesterone receptor is made up of two components, one of which binds to DNA and the other to a histone protein.

6-2 (a) Cells communicate using two general kinds of signals: _____ and
_____ .

(b) Hormones are released into the circulatory system by _____ cells to effect preprogrammed responses in _____ cells.

(c) The _____ exerts primary control over all _____ hormones.

(d) In general, polypeptide hormones are inactivated by _____, and steroids, prostaglandins, and amino acid-derived hormones are inactivated by _____ and _____ .

(e) Since the lack of steroid hormones causes some cells and tissues not to develop properly and others to become incapable of responding to polypeptide and amino acid-derived hormones, steroids sometimes are referred to as _____ hormones.

(f) The molecules that can bind specific hormones and translate that binding directly or indirectly into cellular responses are called _____.

(g) Most hormone receptors are quite asymmetric in shape as is reflected in their large _____.

(h) Steroids pass freely across membranes because of their _____.

(i) _____, _____, _____, and _____ are four general chemical classes of hormones.

(j) A number of hormones are synthesized as _____ and cleaved in the process of secretion.

(k) All steroid hormones are synthesized from _____.

(l) _____ are synthesized from two essential fatty acids.

(m) The intracellular concentration of cAMP reflects the relative activities of _____ and _____ .

(n) Methylxanthines, such as _____ and _____ , inhibit cAMP phosphodiesterase.

(o) cAMP has been called the _____, since it transmits hormonal signals from the cell surface to intracellular components.

(p) Since _____ penetrates cells poorly, a less polar derivative, such as _____, usually is added instead to target cells in tests of the second messenger concept.

(q) In liver cells an increase in cAMP concentration activates _____ by causing dissociation of the _____ subunit, which binds cAMP, from the _____ subunit.

(r) Bidirectionally controlled cellular processes may be regulated by the opposing effects of _____ and _____.

(s) Steroid activation of specific mRNA synthesis is superficially analogous to a _____ controlled _____ bacterial system.

6-3 Histones and nonhistone chromosomal proteins can be dissociated selectively from chromatin and separated from the DNA. Subsequently, the chromatin can be reconstituted by mixing the components and dialyzing very slowly against appropriate buffers. The technique of chromatin reconstitution has been used successfully to determine which component of chromatin is responsible for the difference in binding of 5S hormone-receptor complex to target- and nontarget-cell chromatin. Various reconstituted chromatins from components of chick oviduct and chick erythrocyte are indicated in Table 6-3. Which component of chick oviduct chromatin is responsible for enhanced binding of 5S progesterone-receptor complex?

Table 6-3. Binding of 5S complex to reconstituted chromatin (Problem 6-3). + indicates presence of a component in the reconstitution mixture.

Chick oviduct			Chick erythrocyte			Binding of 5S complex
DNA	Histones	Nonhistones	DNA	Histones	Nonhistones	
+	+	+				High
+	+				+	Low
+		+		+		High
+				+	+	Low
	+	+	+			High
	+		+		+	Low
		+	+	+		High
			+	+	+	Low
Native chick oviduct chromatin						High
Native chick erythrocyte chromatin						Low

6-4 It has been proposed that 5S steroid hormone-receptor complexes induce specific mRNA synthesis by antagonizing specific gene repressors, much as allolactose induces specific synthesis of *lac* operon mRNA by combining with the *lac* repressor. Which of the following observations on hormonal induction are *inconsistent* with negative, inducible control like that of the lactose system.

(a)   Steroid hormones cause increased rates of labeled precursor incorporation into RNA.

(b)   Chromatin isolated from hormone-treated tissues shows increased template activity when used to direct RNA synthesis in cell-free preparations.

(c)   Hormonal induction is inhibited by inhibitors of RNA synthesis.

(d)   New species, or increased concentrations of preexisting species, of RNA appear in hormone-treated tissues.

(e)   Hormone-receptor complexes bind to chromatin.

(f)   Chromosomal "puffing" of specific loci occurs in polytene chromosomes from ecdysone-treated insect larvae.

6-5   A good example of the developmental effects of certain steroid hormones is the regulation of liver glycogen synthetase during fetal development. The glycogen content of fetal rat livers increases greatly between the seventeenth and the nineteenth days of gestation. When explants from the sixteenth day of gestation are placed in organ culture no increase in glycogen content occurs. However, the increase can be induced by addition of hydrocortisone and insulin. On the basis of the data in Table 6-4, propose roles for hydrocortisone and insulin in the increase in glycogen content of rat liver cells during fetal development.

Table 6-4.   Effects of insulin and hydrocortisone on glycogen synthetase and glycogen levels in fetal rat liver cells (Problem 6-5). Explants from 16-day fetuses were incubated for 40 hours with or without hydrocortisone, and then exposed to insulin. Phosphorylated glycogen synthetase (normally inactive) can be assayed in the presence of the allosteric effector glucose-6-P. (From H. J. Eisen *et al., Proc. Natl. Acad. Sci. (U. S.)* **70,** 3455, 1973.)

Hydrocortisone (10 µM)	Insulin (0.1 µ/ml)	Glycogen (mg/g protein)	Nonphosphorylated synthetase (units/g protein)	Total synthetase (units/g protein)
0	0	4	0.10	3.3
0	+	4	0.10	3.0
+	0	5	0.16	14.9
+	+	11	0.48	11.7

6-6   For each of the following liver-cell enzyme defects predict whether the glycogen levels will be high, normal, or low.

(a)   Defective phosphodiesterase

(b)   Protein kinase regulatory subunit that cannot bind cAMP

(c)   Defective insulin receptor

(d)   Defective adenyl cyclase

(e)   Protein kinase regulatory subunit that cannot bind to the catalytic subunit

(f)   Defective phosphorylase kinase

(g)   Defective protein phosphatases

6-7   When two or more structurally different hormones stimulate adenyl cyclase in the same target cell, the maximally effective hormone concentrations are not additive. Instead, the activity of adenyl cyclase reaches only the level that is attained with the most effective hormone. Given this observation, which, if any, of the following models for the arrangement of regulatory and catalytic subunits in the adenyl cyclase complex are ruled out.

(a)   Individual regulatory subunits attached to different kinds of catalytic subunits [Figure 6-8(a)].

(b)   Individual regulatory subunits attached to one kind of catalytic subunit [Figure 6-8(b)].

(c)   Multiple regulatory subunits attached to one catalytic subunit [Figure 6-8(c)].

(d)   One regulatory subunit with multiple binding sites attached to one catalytic subunit [Figure 6-8(d)].

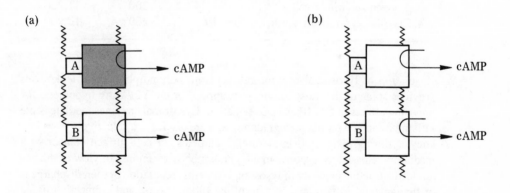

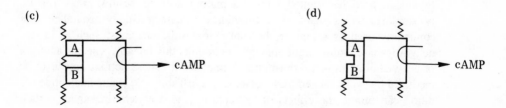

Figure 6-8. Models for the adenyl cyclase complex (Problem 6-7).

6-8 The cellular localization of the receptors for two hormones (one steroid-one polypeptide) have been examined as described in Table 6-5.
(a) Describe the cellular location of the receptors as precisely as the data allow.
(b) From your general knowledge of hormone action, decide which hormone most likely is the polypeptide and which is the steroid.

Table 6-5. Experimental results on the cellular localization of hormone receptors (Problem 6-8). Trypsin treatment releases a number of peptides from the cell surface but does not affect proteins on the inside of a cell. A and B are two different hormones.

Cell treatment	Labeled hormone bound to cell or cell homogenate (cpm/mg protein)	
	A	B
1. None (whole cells)	9,000	11,000
2. Homogenize	8,700	10,900
3. Trypsinize (whole cells)	200	11,200
4. Trypsinize; remove trypsin; homogenize	200	10,900

6-9 Regulation of lysosomal enzyme release from polymorphonuclear leukocytes provides some of the most convincing support for the Yin/Yang hypothesis for opposing effects of cAMP and cGMP action. Lysosomal enzymes mediate acute inflammation and cartilage degradation and are discharged selectively from leukocytes during phagocytosis of foreign particles, or cell-surface contact with mediators released in response to antigen-antibody reactions. In view of the potential destructive capacity of lysosomal contents, regulation of their discharge is of fundamental importance in controlling inflammation and connective tissue destruction.

When human leukocytes are treated with specially prepared particles (zymosan particles treated with rheumatoid arthritis serum), they release lysosomal enzymes such as β-glucuronidase, a characteristic lysosomal enzyme commonly used as a marker, but not cytoplasmic enzymes, such as lactate dehydrogenase. Addition of epinephrine inhibits this release, whereas addition of acetylcholine (a neurotransmitter) stimulates it. These actions appear to be mediated by cyclic nucleotides. As indicated in Table 6-6, cyclic nucleotide addition can mimic the effects of epinephrine and acetylcholine under certain conditions.
(a) Which cyclic nucleotide appears to be elevated by epinephrine and which by acetylcholine?
(b) Explain the results for Lines 3, 4, and 5 of Table 6-6.

Table 6-6. Effects of cAMP and cGMP on release of β-glucuronidase from human leukocytes (Problem 6-9). (Adapted from L. J. Ignarro and W. J. George, *Proc. Natl. Acad. Sci. (U. S.)* **71**, 2027, 1974.)

Experimental condition[a]	Release of β-glucuronidase activity (arbitrary units)
1. Control release	$(52.2 \pm 3.1)^b$
2. Total enzyme activity in cells	$(257.4 \pm 15.8)^b$
3. Cyclic AMP	45.9 ± 4.0
4. Cyclic AMP + 1 μM theophylline	31.8 ± 3.1^c
5. Dibutyryl cyclic AMP	29.8 ± 3.6^c
6. Cyclic GMP	119.2 ± 12.4^c
7. Dibutyryl cyclic GMP	139.4 ± 15.0^c
8. 8-Bromocyclic GMP	160.8 ± 12.6^c

[a]Human leukocytes (5×10^6) were incubated with or without compound(s) for 5 min. at 37°C before addition of rheumatoid arthritic serum-treated zymosan particles, and then incubated for an additional 15 min. at 37°C. Data represent the mean $\pm$ standard error of four separate determinations.
[b]No compound was added.
[c]Significantly different ($p < 0.01$) from control enzyme release, by student's t-test.

6-10   (a)   Briefly outline three possible mechanisms by which cAMP and cGMP could produce opposing effects as demanded by the Yin/Yang hypothesis. Assume that all actions of cAMP are mediated through protein kinases.
(b)   In which, if any, of these hypothetical mechanisms would the concentrations of cAMP and cGMP be expected to vary inversely with one another?

6-11   cAMP has been implicated in the regulation of a variety of membrane-associated functions, including contact inhibition, hormone secretion, cellular permeability, and synaptic transmission. In Table 6-7 the distribution of a protein kinase in human erythrocytes is compared with that of three reference enzymes. In Table 6-8 the ability of the protein kinase to phosphorylate membrane proteins upon addition of cAMP is examined in the same sets of preparations.
(a)   What is the location of the protein kinase in the membrane?
(b)   Based on these results, propose a mechanism whereby a hormone might alter the rate of a membrane-associated function, such as cellular permeability.

6-12   Upon stimulation of chick oviduct with estradiol there is a 10^5-fold increase in the total number of ovalbumin mRNA molecules present in the cell. There are two general ways in which such an amplification might occur: (1) The number of RNA transcripts from each ovalbumin gene might increase or (2) the number of ovalbumin genes might increase.

      The number of copies of specific genes can be determined by reassociation kinetics. The specific gene "probe" that is needed for such studies can be made

Table 6-7.   Accessibility of protein kinase and reference enzymes in human erythrocyte membrane preparations (Problem 6-11). Acetylcholinesterase is a marker for the outside of the membrane, whereas glyceraldehyde-3-phosphate dehydrogenase and diaphorase are markers for the inside of the membrane. Empty cell membranes are called ghosts due to their appearance in the electron microscope. Protein kinase was assayed by addition of protamine, an external substrate. (From C. S. Rubin *et al., Proc. Natl. Acad. Sci. (U. S.)* **70,** 3735, 1973.)

| | Percent of total activity | | | |
	Acetylcholine-esterase	Glyceraldehyde-3-phosphate dehydrogenase	Diaphorase	Protein kinase
Permeable ghosts	100	99	90	91
Sealed right-side-out ghosts	106	14	12	19
Sealed inside-out vesicles	22	98	88	87

Table 6-8.   Phosphorylation of two endogenous erythrocyte membrane proteins (Problem 6-11). ^{32}P incorporation is measured in the presence of externally added cAMP. (From C. S. Rubin *et al., Proc. Natl. Acad. Sci. (U. S.)* **70,** 3735, 1973.)

| | ^{32}P-incorporated (pmoles) | |
	Protein I	Protein II
Permeable ghosts	4.5	7.8
Sealed right-side-out ghosts	0.9	1.2
Sealed inside-out vesicles	6.0	7.3

by purifying the specific mRNA (generally very difficult), and subsequently copying it with RNA-directed DNA polymerase to produce a complementary DNA (cDNA) copy of the RNA [Essential Concept 5-8(a)]. The resulting cDNA typically has a very high specific radioactivity of 5×10^7 to 10×10^7

cpm/μg, which is required to detect genes that are present as single copies.

The purification of ovalbumin mRNA from chick oviduct is simplified because it is the major mRNA species in the cell (approximately 65% of the actively synthesized protein in a stimulated oviduct is ovalbumin). The cDNA made from this mRNA can be shown to be a faithful copy of the RNA by its complete hybridization to the RNA. The reassociation curve for total DNA from stimulated chick oviduct is shown in Figure 6-9.

(a)   Sketch the curves for association of cDNA with the oviduct DNA assuming there are 0, 1, or 100 ovalbumin genes per haploid amount of cell DNA.

(b)   The actual $C_0t_{\frac{1}{2}}$ for cDNA association in this experiment was 480. Which of the two proposed mechanisms is responsible for the increase in ovalbumin mRNA in the cell?

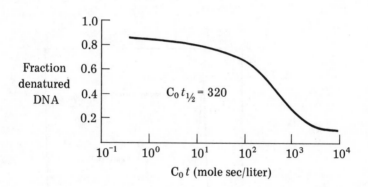

Figure 6-9. Reassociation kinetics of total DNA from stimulated chick oviduct (Problem 6-12). (Adapted from S. E. Harris *et al., Proc. Natl. Acad. Sci. (U. S.)* **70,** 3776, 1973.)

6-13   One line of cultured mouse lymphoma cells, S49.1 TB4, is quite sensitive to cAMP. Elevated internal levels inhibit their growth and ultimately lead to cell death. As expected, these cells are also sensitive to added dibutyryl cAMP. However, resistant cells can be selected from the predominantly sensitive population by growth in the presence of dibutyryl cAMP. Such cells are resistant not only to dibutyryl cAMP, but also to the prostaglandin PGE$_1$, which kills sensitive cells by increasing intracellular cAMP levels. The binding of cAMP to its receptors in sensitive and resistant cell homogenates has been examined as shown in Figure 6-10. Deduce which, if any, of the following three explanations for resistance is consistent with all the data.

(1) The transport system for dibutyryl cAMP is defective in resistant cells. cAMP binds to this transport protein in sensitive cell homogenates and therefore does not bind in resistant cell homogenates.

(2) The regulatory subunit of a protein kinase is defective in resistant cells. cAMP binds to this regulatory subunit in sensitive cell homogenates and therefore does not bind in resistant cell homogenates.

(3) Resistant cells contain a new activity that is capable of modifying cAMP so that it no longer can bind to its receptor. The ''free cAMP'' in resistant cell homogenates is actually a modified cAMP that elutes in approximately the same way as cAMP.

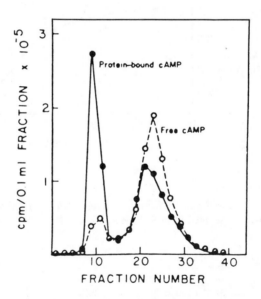

Figure 6-10. Gel filtration of ^{3}H-cAMP after incubation with cell homogenates from sensitive and resistant cells (Problem 6-13). Cells were homogenized and particulate material was removed by centrifugation. The resulting preparations from sensitive (●) and resistant (○) cells were incubated with ^{3}H-cAMP in the presence of $5mM$ theophylline. If a mixture of the two extracts is incubated with ^{3}H-cAMP, the gel filtration pattern observed is the same as that for the sensitive cell extract. (From V. Daniel *et al., Proc. Natl. Acad. Sci. (U. S.)* **70,** 76, 1973.)

6-14 A hypothetical polypeptide hormone binds to its receptor with a rate constant for association (k_a) of $3.0 \times 10^4 M^{-1} \sec^{-1}$ and a rate constant for dissociation (k_d) of $6.9 \times 10^{-6} \sec^{-1}$.

(a) Calculate the equilibrium constant for the dissociation reaction.

(b) Calculate the half-life for the hormone-receptor complex. Since dissociation of the complex is a first-order reaction, the fraction of complex (HR) remaining at any time t is given by the expression

$$\frac{[HR]}{[HR]_0} = e^{-k_d t}$$

in which $[HR]_0$ is the original concentration of the complex.

(c) Would you expect the magnitude of target-cell response to this hormone to be related closely to the concentration of circulating hormone?

(d) Would you expect to find any real hormones with these properties?

6-15 The concentration of a hormone (H) in the circulation is controlled by a balance between a synthetic reaction that produces it from a precursor (P) and a degradative reaction that converts it to a degradation product (D). The opposing reactions can be represented as

$$P \xrightarrow{k_s} H \xrightarrow{k_b} D$$

in which k_s and k_b are rate constants for synthesis and breakdown respectively. The rate of hormone production will be equal to $[P]k_s$, and the rate of hormone disappearance will be equal to $[H]k_b$.

(a) If [P] is increased quickly from zero to a finite level and maintained constant, do you expect that the concentration of circulating hormone will (1) continue to increase until [P] is again decreased, (2) increase to a plateau and remain constant, or (3) increase to a maximum, decrease to a minimum, and continue to oscillate between these two levels?

(b) From the expressions for rates of production and breakdown of H, derive an expression for [H] as a function of time (t) under conditions where [P], k_s, and k_b are constant, and the initial value of [H] = 0.

(c) The value of k_s for many hormones probably is determined by the catalytic properties of a regulatory enzyme that is responsive to changes in concentrations of metabolites or other hormones. For a hormone whose k_s is variable, what will be the effect on the steady-state level of circulating H if k_s is increased by a factor of 2?

6-16 Steroid hormones (S) enter target cells and bind to cytoplasmic receptor proteins (R) with high affinity. For a given hormone, there are about 10^4 R molecules per cell. The dissociation constant for the SR complex is

$$K_{SR} = \frac{[S][R]}{[SR]} \simeq 10^{-10} M \tag{6-1}$$

SR complexes enter the nucleus and bind to chromatin to somehow stimulate synthesis of specific mRNA's. Two questions of current interest are: How many specific binding sites are there per cell nucleus, and what are the dissociation constants for SR binding to these sites? To confuse the issue, SR complexes bind weakly and nonspecifically to all chromatin and DNA preparations to form SRDNA complexes with the dissociation constant

$$K_{SRDNA} = \frac{[SR][DNA]}{[SRDNA]} \simeq 0.3 \,\text{mg/ml DNA} \qquad (6\text{-}2)$$

in which [DNA] is expressed as mg/ml, and the other two concentrations are in moles/liter. The nonspecific binding presumably is not related directly to the physiological effects of the hormone. Assume that the volume of a target cell is 2×10^{-9} ml, and that its haploid DNA content is 3×10^{-12} g. Physiological effects are first observed *in vivo* at circulating steroid hormone concentrations of about $10^{-9}M$ ($[S] = 10^{-9}M$).

(a)   What do you predict would be the dissociation constant of the specific DNA sites (SS) for SR binding?

(b)   Suppose that you wish to demonstrate the existence of specific sites by comparing the binding of radioactively labeled SR complex to nuclei of equal DNA content from target cells and from an unrelated organism that shows no steroid response. If your assay allows you to detect an increment of 10% over the background of nonspecific binding, how many specific binding sites must be present per target cell nucleus for you to be able to detect them?

ANSWERS

6-1   (a)   False. The term hormones is restricted to chemical messages that are transmitted via the circulatory system. For example, neuronal transmitters, which communicate from presynaptic to postsynaptic nerve cells across the synapses, are not classed as hormones.

(b)   True

(c)   False. The adenohypophesis is not under central nervous system control.

(d)   False. Only the polypeptide and amino acid-derived hormones are stored in secretory granules.

(e)   True

(f)   True

(g)   False. Steroid receptors are located in the cytoplasm.

(h)   False. The insulin binding site faces outward.

(i)   False. Neither hormones nor their receptors are inactivated by binding.

(j)   True

(k)   True

(l)   True

(m)   False. cAMP is used only as an allosteric effector.

(n)   True

(o)   False. Phosphodiesterase does not appear to be regulated. Adenyl cyclase consists of separate subunits for hormone recognition and cAMP synthesis.

(p)   True

(q)   False. Insulin lowers cAMP levels, whereas theophylline raises cAMP levels.

(r)   False. Cortisol is a steroid hormone and does not use cAMP as a second messenger.

(s)   True

(t)   True

(u)   False. One component of the progesterone receptor binds to DNA and the other component binds to a *nonhistone* chromosomal protein.

6-2   (a)   electrical, chemical

        (b)   endocrine, target

        (c)   hypothalamus, pituitary (or hypophysial)

        (d)   proteolysis, modification, degradation

        (e)   developmental

        (f)   hormone receptors

        (g)   axial ratios

        (h)   hydrophobicity (or nonpolar character)

        (i)   Polypeptides, amino acid derivatives, steroids, prostaglandins

        (j)   prohormones

        (k)   cholesterol

        (l)   Prostaglandins

        (m)   adenyl cyclase, cAMP phosphodiesterase

        (n)   theophylline, caffeine

        (o)   second messenger

        (p)   cAMP, dibutyryl cAMP

        (q)   protein kinase, regulatory, catalytic

        (r)   cAMP, cGMP

        (s)   positively, inducible

6-3   Enhanced binding of the 5S complex is correlated in all cases with the presence of nonhistone chromosomal proteins from the target cell. Presumably the enhanced binding is due to one or a few nonhistone chromosomal proteins. It is not clear whether the enhanced binding results from novel species of nonhistone proteins that are present only in the target cells or from proteins that are present in target cells in increased amounts.

6-4   (e)   Hormone-receptor complexes, if they act in a manner analogous to that of allolactose, should not be found associated with chromatin. If the hormone-receptor complex must bind to chromatin to function, positive rather than negative control is implicated.

6-5   The data in Table 6-4 indicate that there are two controls on glycogen synthetase. Hydrocortisone increases the total amount of glycogen synthetase, whereas insulin increases the fraction that is in the active form. As indicated in the fourth line of the table, both hormones are necessary for a significant increase in glycogen content.

6-6   (a)   Low
       (b)   High
       (c)   Low to normal
       (d)   High
       (e)   Low
       (f)   High
       (g)   Low

6-7   Models (a) and (b) are ruled out. Since the regulatory subunits control different sets of adenyl cyclases, these models predict that the complexes should behave independently, thereby producing additive responses to mixtures of hormones.

6-8   (a)   The receptor for hormone A must be a protein that is located on the cell surface, since it is sensitive to trypsin. Furthermore, the absence of hormone binding by the cell preparation that had been trypsinized and then homogenized indicates that the receptors are confined solely to the cell surface; that is, there must be few, if any, receptors free in the cytoplasm or in intracellular membranes.
       The cellular localization of the receptor for hormone B is not fixed by these studies. The only limitation on the nature of this receptor is that if it is on the cell surface, it must be insensitive to trypsin.
       (b)   The results in Table 6-5 are consistent with hormone A being the polypeptide and hormone B being the steroid. The results shown for hormone A are similar to actual data for insulin binding to its receptor.

6-9   (a)   The effects of cAMP and epinephrine correlate, as do the effects of cGMP and acetylcholine. Thus it seems most likely that epinephrine increases intracellular cAMP levels and that acetylcholine increases intracellular cGMP levels. Direct measurements (not shown) of changes in intracellular cyclic nucleotide concentration after stimulation by these compounds verify this conclusion.
       (b)   The results in Lines 3, 4, and 5 of Table 6-6 are understandable given that cAMP penetrates cells poorly, and that theophylline inhibits phosphodiesterase. Addition of cAMP (Line 3) shows no significant difference from the control because it does not get into the cell. The decrease in β-glucuronidase release after addition of cAMP and theophylline (Line 4) is due solely to the theophylline, which inhibits intracellular phosphodiesterase and thereby causes an increase in cAMP concentration. Addition of dibutyryl cAMP (Line 5) causes a decrease in enzyme release because this analogue can penetrate the cell.

6-10   (a)   The possible mechanisms by which cGMP could produce an opposite effect to cAMP fall into three general classes:

(1)   cGMP could affect cAMP levels directly by inhibiting adenyl cyclase or activating cAMP phosphodiesterase.

(2)   cGMP could reverse the effects of cAMP by inactivating the cAMP-activated protein kinase, or by activating a phosphatase that would dephosphorylate phosphorylated substrates of protein kinase.

(3)   cGMP could act independently of the cAMP system, through a separate set of enzymes, to produce opposing effects.

(b)   The concentrations of cAMP and cGMP would be expected to vary inversely with one another in none of the mechanisms listed in Part (a). In mechanism (1) the concentration of cAMP would decrease or increase as the concentration of cGMP changed. However, the concentration of cGMP would not be expected to change with increases or decreases in cAMP concentration. The concentrations of cAMP and cGMP would be expected to vary inversely with one another only if each had an effect on the other's synthesis or degradation.

6-11   (a)   Since the protein kinase shows the same distribution as the two markers for the inside of the membrane and is activated by cAMP only in permeable ghosts and inside-out vesicles, it must be located on the inside of the membrane.

(b)   One possible mechanism for hormone alteration in the rate of a membrane-associated function, such as cellular permeability, is as follows: The hormone interacts with a surface receptor to activate adenyl cyclase, thereby causing an intracellular increase in cAMP concentration. Increased cAMP activates the membrane-associated protein kinase, which in turn phosphorylates specific transport proteins. Depending on whether cellular permeability were to be increased or decreased, the phosphorylated transport protein would have increased or decreased activity.

6-12   (a)   See Figure 6-11. If there were no ovalbumin genes, the cDNA could not

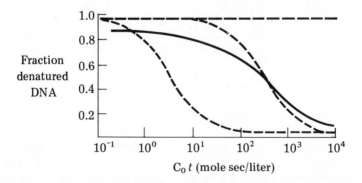

Figure 6-11. Reassociation kinetics of labeled ovalbumin message cDNA (dashed lines) with total DNA from chick oviduct containing 0, 1, and 100 ovalbumin genes (Answer 6-12). Solid line indicates the reassociation kinetics of total DNA from stimulated chick oviduct for reference.

associate and would have an infinite $C_0t_{\frac{1}{2}}$. If there were one ovalbumin gene, the cDNA would associate at the same rate as chick cell unique DNA, so that $C_0t_{\frac{1}{2}} = 320$. If there were 10^2 ovalbumin genes, the cDNA would associate at 10^2 times the rate of unique DNA, so that $C_0t_{\frac{1}{2}} = 3.2$.

(b) Since the $C_0t_{\frac{1}{2}}$ for cDNA is approximately that of unique DNA, there is most likely one ovalbumin gene (two per diploid amount of DNA) whose transcription is increased enormously by estradiol stimulation.

6-13 Only Explanation (2) is consistant with all the data. If a transport system for dibutyryl cAMP were defective, the resistant cells still would be sensitive to PGE_1. (In addition, the transport protein probably would be removed from the homogenates with the particulate matter. Moreover, dibutyryl cAMP probably does not require a transport system to cross the membrane.) If there were a cAMP-modifying activity in resistant cells, the mixture of extracts would have yielded a gel-filtration pattern like that for the resistant cells rather than that for the sensitive cells.

6-14 (a) The equilibrium constant for dissociation of the hormone-receptor complex is equal to the ratio of the forward and reverse rate constants:

$$K_{eq} = \frac{k_d}{k_a} = \frac{6.9 \times 10^{-6} \text{ sec}^{-1}}{3.0 \times 10^4 M^{-1} \text{ sec}^{-1}}$$

$$K_{eq} = 2.3 \times 10^{-10} M$$

(b)
$$\frac{[HR]}{[HR]_0} = e^{-k_d t}$$

$$2.3 \log \frac{[HR]}{[HR]_0} = -k_d t$$

At the end of one half-life, $[HR]/[HR]_0 = 0.5$; therefore

$$t_{1/2} = \frac{-2.3 \log(0.5)}{6.9 \times 10^{-6} \text{ sec}^{-1}}$$

$$t_{1/2} = \frac{(-2.3)(-0.3)}{6.9 \times 10^{-6} \text{ sec}^{-1}}$$

$$t_{1/2} = 10^5 \text{ sec} \simeq 28 \text{ hr}$$

(c) The target-cell response to this hormone will not be related closely to the concentration of circulating hormone. After more than a day, half the original hormone that was bound to the receptor will remain bound regardless of the circulatory concentration.

(d) It is unlikely that any real hormones with these properties will be found. Insensitivity to circulating hormone concentration is inconsistent with the functions of known polypeptide hormones, which allow efficient short-term control of metabolism in target tissues.

6-15   (a)   As long as [P] remains constant, H will be produced at a constant rate. However, the rate of H degradation, which is proportional to [H], will increase as [H] increases, until $[P]k_s = [H]k_b$. From this point on, [H] will remain constant at the steady-state level:

$$[H] = [P]\frac{k_s}{k_b} \tag{6-3}$$

This expression is derived in Part (b).

(b)

$$\frac{d[H]}{dt} = [P]k_s - [H]k_b$$

Separation of variables for integration gives

$$\int_0^{[H]} \frac{d[H]}{[P]k_s - [H]k_b} = \int_0^t dt$$

From a table of integrals,

$$-\frac{1}{k_b} \ln([P]k_s - [H]k_b)\Big|_0^{[H]} = t$$

or

$$\ln([P]k_s - [H]k_b) - \ln([P]k_s) = -k_b t$$

Collection of terms and conversion to exponential form gives

$$1 - \frac{[H]k_b}{[P]k_s} = e^{-k_b t}$$

which rearranges to

$$[H] = [P]\frac{k_s}{k_b}(1 - e^{-k_b t}) \tag{6-4}$$

As t increases, the second term of Equation 6-4 tends to zero, yielding Equation 6-3 for the steady-state value of [H].

(c)   From Equation 6-3, the steady-state value of [H] is directly proportional to k_s. Therefore [H] will double if k_s is increased by a factor of 2.

6-16   (a)   To compete effectively for the SR complex in the presence of nonspecific binding, the dissociation constant for the specific binding sites must be roughly equal to the molar concentration of free SR complex at free steroid concentrations of $\sim 10^{-9}M$. The free SR concentration may be calculated as follows. The total concentration of receptor in the cell is about

$$\frac{(10^4 \text{ molecules/cell})(1000 \text{ ml/liter})}{(2 \times 10^{-9} \text{ ml/cell})(6 \times 10^{23} \text{ molecules/mole})} = 10^{-8}M$$

The ratio of [R] to [SR] at a steriod concentration of $10^{-9}M$ can be calculated from Equation 6-1:

$$\frac{[R]}{[SR]} = \frac{K_{SR}}{[S]} = \frac{10^{-10}}{10^{-9}} = 0.1$$

that is, about 90% of the receptor molecules are complexed, so that $[SR] \simeq 10^{-8}M$.

The concentration of DNA in the cell is about

$$\frac{6 \times 10^{-9} \text{ mg DNA per diploid nucleus}}{2 \times 10^{-9} \text{ ml of cell volume}} = 3 \text{ mg/ml}$$

From this concentration, the ratio of [SRDNA] to $[SR]_{free}$ can be calculated using Equation 6-2:

$$\frac{[SRDNA]}{[SR]_{free}} = \frac{[DNA]}{K_{SRDNA}} = \frac{3 \text{ mg/ml}}{0.3 \text{ mg/ml}} = 10$$

that is, about 90% of the total SR complex is bound to DNA, so that the concentration of free SR is somewhat less than $10^{-9}M$. Therefore the dissociation constant of the specific sites for DNA binding, K_{SRSS}, must be about $10^{-9}M$.
(b)   To be detected, the amount of SR bound at specific sites must be at least 10% of the amount bound nonspecifically or

$$0.10[SRDNA] = [SRSS] \tag{6-5}$$

If the dissociation constant for specific sites is taken to be $10^{-9}M$, then

$$K_{SRSS} = 10^{-9} = \frac{[SS][SR]_{free}}{[SRSS]} \tag{6-6}$$

in which [SS] and [SRSS] represent the concentrations of free and occupied specific binding sites, respectively. Substitution from Equations 6-2 and 6-6 into Equation 6-5 gives

$$(0.1)\frac{[DNA][SR]}{0.3 \text{ mg/ml}} = \frac{[SS][SR]}{10^{-9}M}$$

or

$$\frac{[SS]}{[DNA]} = 3.3 \times 10^{-10} \text{ mole specific sites/ g DNA}$$

Since a diploid nucleus contains 6×10^{-12} g of DNA, the number of sites per nucleus must be at least

$$(6 \times 10^{23} \text{ sites/mole})(3.3 \times 10^{-10} \text{ mole/g})(6 \times 10^{-12} \text{ g/nucleus})$$
$$= 1000 \text{ sites/nucleus}$$

7 Molecular Evolution

Proteins and nucleic acids carry in their amino acid or nucleotide sequences clues to their own evolutionary history and, collectively, to the evolutionary history of the organism itself. These clues can be deciphered by comparing the sequences and conformations of contemporary proteins or genes. Ultimately, such studies should lead to an understanding of the molecular mechanisms responsible for evolution and make possible the construction of detailed evolutionary histories (genealogical trees) that describe the origin and history of many living species. This chapter considers two primary aspects of molecular evolution: the genetic mechanisms by which gene variation is produced and the means by which such variation subsequently is fixed into a population of organisms.*

ESSENTIAL CONCEPTS

7-1 All living organisms probably descended from a common ancestor

(a)   Although the origins of life are uncertain, the general outline of organic evolution can be inferred from biochemical, biological, geological, and fossil evidence (Figure 7-1). From an inorganic environment, evolution sequentially generated simple organic molecules, including amino acids and nucleotides; polymers with primitive catalytic and reproductive capabilities; primitive cells containing these polymers; a proto-organism that synthesized proteins to control metabolism, growth, and replication; and finally bacteria and higher organisms.

Throughout evolution many now-extinct organisms arose as indicated by the dashed lines in Figure 7-1. It is important to remember that contemporary biologists can examine only the components of successful evolutionary lines.

(b)   The basic similarity of all cellular metabolism suggests that present-day organisms arose from a common ancestor, or proto-organism. Shared features, such as utilization of ATP for energy transfer and short-term storage, pathways and cofactors for synthesis and degradation of many cell constituents, a common set of L-amino acids, and the genetic code, probably could not have arisen independently in two or more ancestral organisms. An important consequence of common ancestry is that all contemporary genes must have been derived from the primordial genes of the proto-organism.

---

*This chapter benefited greatly from the ideas of Russell F. Doolittle and is based in part on his forthcoming review "Protein Evolution," in H. Neurath and R. Hill (Ed.) *The Proteins* (Academic Press, New York, in press, 1976). We are grateful for his generosity in providing us with this article in advance of its publication.

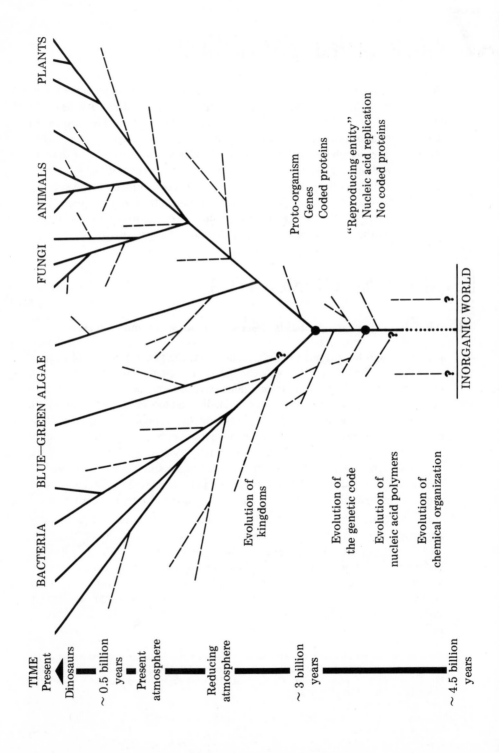

◀ **Figure 7-1.** The phylogenetic tree of life. [Adapted from M. O. Dayhoff and R. V. Eck, "Tracing biochemical evolution," in M. O. Dayhoff (Ed.), *Atlas of Protein Sequence and Structure* Vol. 5, National Biomedical Research Foundation, Washington, D.C., 1972.]

7-2 Evolution depends upon the generation of variant genes and their subsequent fixation in the population

(a) The variant genes that appear to be most important in evolution are generated by single-base substitutions and gene duplications. Other kinds of genetic alteration, such as frameshifts and deletions, seem to be less important. The extra gene copies produced by gene duplication are the "raw material" for evolution since they are free to diverge from their original sequences and in time can code for proteins with new functions.

(b) If a variant gene is to be evolutionarily important, it must spread throughout the population to become the predominant form. The means by which variant genes become *fixed* in a population are considered in Essential Concept 7-6.

7-3 Mutation rate and genetic code structure restrict the possible amino acid substitutions

(a) The rate at which amino acid replacements are made depends in part upon the rate of mutation. Spontaneous mutation rates are affected by a variety of external factors such as dietary mutagens, temperature, and radiation exposure, and internal factors such as the fidelity of an organism's own replication enzymes. One mutant *E. coli* DNA polymerase has been shown to increase the spontaneous mutation rate by 100 to 1000 fold. These variables make a precise determination of mutation rate impossible. However, a number of experiments suggest an average mutation rate for single-base replacements of about 10^{-10} substitutions per nucleotide per DNA replication. Given this mutation rate the probability in each replication cycle of a single substitution in a codon is about 3×10^{-10}, and the probability of two substitutions in a single codon is 9×10^{-20}. Consequently, single-base substitutions in a codon are much more likely than are two- and three-base substitutions.

(b) The genetic code is composed of 61 codons for amino acids and three codons for chain termination (Figure 7-2). Since each amino acid codon can be converted to any of nine different codons by single-nucleotide substitutions (3 alternatives at each of the 3 nucleotides), $61 \times 9 = 549$ distinct single-nucleotide substitutions are possible. The resulting codons can code either for the same amino acid (*silent* mutation), for a different amino acid (*missense* mutation), or

First position (5′ end)	Second position				Third position (3′ end)
	U	C	A	G	
U	Phe	Ser	Tyr	Cys	U
	Phe	Ser	Tyr	Cys	C
	Leu	Ser	Term	Term	A
	Leu	Ser	Term	Trp	G
C	Leu	Pro	His	Arg	U
	Leu	Pro	His	Arg	C
	Leu	Pro	Gln	Arg	A
	Leu	Pro	Gln	Arg	G
A	Ilu	Thr	Asn	Ser	U
	Ilu	Thr	Asn	Ser	C
	Ilu	Thr	Lys	Arg	A
	Met	Thr	Lys	Arg	G
G	Val	Ala	Asp	Gly	U
	Val	Ala	Asp	Gly	C
	Val	Ala	Glu	Gly	A
	Val	Ala	Glu	Gly	G

Figure 7-2. The genetic code.

for chain termination (*nonsense* mutation) (Table 7-1). There are $20 \times 19/2 = 190$ possible interchanges for the twenty amino acids. Of these interchanges, only 75 can result from single-base changes (101 of them require two-base changes, and 14 of them require three-base changes).

(c)   Analysis of variant proteins that have not yet been acted on by the selective forces of evolution confirms that amino acid replacements arise by single-base substitution. All of the more than 200 amino acid replacements that have been observed in the most thoroughly studied group of variant proteins, the abnormal human hemoglobins, have arisen by single-base substitution according to the code.

Table 7-1.   Types of single nucleotide substitutions.

Type	Fraction of Total		
Silent	134/549	or	24.4%
Missense	392/549	or	71.4%
Nonsense	23/549	or	4.2%

7-4 Conservative amino acid substitutions are fixed most frequently during the evolution of proteins

(a) Amino acid side chains can be grouped according to their physical and chemical properties, that is, their size, shape, number of atoms, functional groups, pK_a's of ionization, and degree of polar or nonpolar character. One such grouping is shown in Figure 7-3. Note that it is difficult to decide on a best

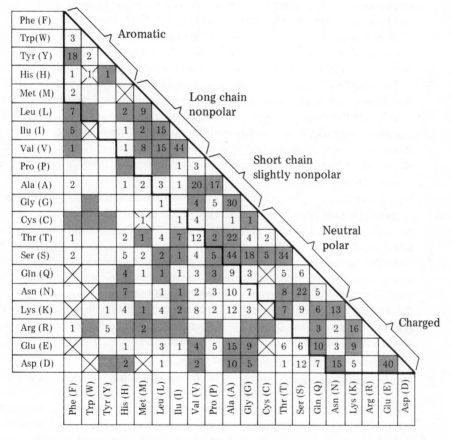

Figure 7-3. Matrix of possible amino acid interchanges. Amino acids are arranged along the axes according to their degree of relatedness. Gray, white, and crossed squares represent interchanges that are possible from one, two, and three base substitutions, respectively. Observed amino acid replacements are indicated by numbers. [Adapted from R. Doolittle, "Protein Evolution," in H. Neurath and R. Hill (Ed.), *The Proteins* (Academic Press, New York, in press). Data from M. O. Dayhoff, R. V. Eck, and C. M. Park, "A model of evolutionary change in proteins," in M. O. Dayhoff (Ed.), *Atlas of Protein Sequence and Structure* Vol. 5, National Biomedical Research Foundation, Washington, D.C., 1972.]

grouping since almost every amino acid side chain has some property by which it resembles side chains of other groups (for example, His could be grouped with the charged amino acids). *Conservative* substitutions are defined as those that involve amino acids with similar properties (e.g., Arg/Lys, Phe/Tyr, Glu/Asp). *Radical* substitutions are defined as those that involve amino acids with markedly different side chain character (e.g., Asp/Val, Phe/Ser, Arg/Ilu).

(b) The structure of the genetic code imparts a bias in favor of conservative amino acid substitutions (WWBH Essential Concept 19-2). This bias can be demonstrated by analysis of a 20×20 matrix, in which the amino acids are ordered along the axes according to their relative homology (Figure 7-3). In such a matrix conservative amino acid interchanges will be close to the diagonal, whereas less conservative and radical interchanges will be farther away from the diagonal. If the heavy line in Figure 7-3 is used arbitrarily to distinguish between conservative and nonconservative amino acid interchanges, then only 70 (37%) of the possible 190 amino acid interchanges would be classified as conservative. If all base substitutions are assumed to be equally probable and if each amino acid interchange that arises by single-base substitution (gray squares in Figure 7-3) is weighted according to the number of different ways it can arise (e.g., the interchanges Val/Met, Val/Leu, and Val/Ilu can occur in 1, 6, and 3 different ways, respectively), then 53% of all single-base substitutions will produce conservative amino acid interchanges.

(c) Amino acid replacements that occur during evolution tend to be more conservative than expected even allowing for the bias of the genetic code. Seven hundred ninety of the amino acid replacements that have been observed among members of several groups of related proteins (e.g., α hemoglobins, β hemoglobins, and cytochrome c) are indicated by the numbers in Figure 7-3. Sixty-eight percent of these replacements are conservative, whereas only 53% would be expected to be conservative on the basis of the restrictions imposed by the structure of the genetic code. Thus natural selection favors amino acid replacements that are primarily conservative, although the definition of conservative and radical is to some extent arbitrary.

7-5 Duplicated genes can evolve to code for proteins with new functions

(a) Gene duplications range in size from fractions of a gene, to whole genes, to microscopically visible pieces of chromosomes, to the entire genome. *Polyploidization,* as genome duplication is called, is a common occurrence among plants but is uncommon among animals, although it has been observed in fish and amphibia. Apparently the smaller-scale, gene-sized duplications are most important in evolution.

(b) Though the genetic mechanisms underlying small-scale gene duplication are unknown, unequal crossing-over probably plays an important role. Unequal crossing-over is a recombination event that occurs in a misaligned region between homologous chromosomes (Figure 7-4). Misalignment presumably is due

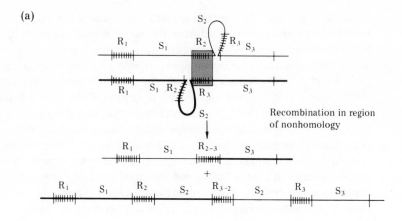

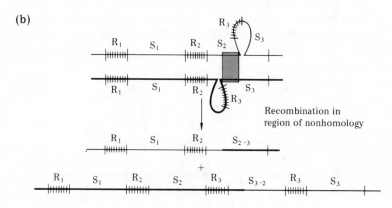

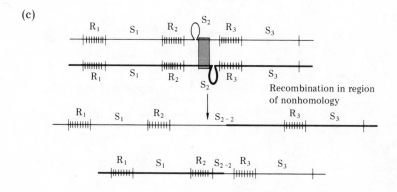

Figure 7-4. Unequal crossing-over leads to duplicated or hybrid genes. (a) Discrete gene duplication. (b) Hybrid gene formation. (c) Contiguous gene duplication. R_x indicates repetitive sequence elements, whereas S_x indicates structural genes. Each line indicates a chromosomal duplex.

to pairing between DNA sequences with only partial homology. Such non-homologous pairing could result from similar repetitive sequences interspersed among structural genes, similar sequences in nonhomologous structural genes, or similar sequences in nonhomologous regions of a single structural gene.

If both crossover points are outside the structural gene, one of the resulting chromosomes will contain two identical copies of the gene [*discrete* gene duplication, Figure 7-4(a)]. If one of the crossover points is inside the structural gene, one of the resulting chromosomes will contain a portion of the structural gene fused to another DNA sequence [*hybrid* gene formation, Figure 7-4(b)]. If both crossover points are inside the structural gene, one of the resulting chromosomes will contain an altered structural gene that contains an internal duplication [*contiguous* gene duplication, Figure 7-4(c)]. In each case the DNA that is added to one chromosome is deleted from the other.

(c)   Duplicated genes rapidly can increase or decrease in number by recombination in misaligned regions that result from *homologous* pairing, as shown in Figure 7-5. As a family of homologous genes increases in number,

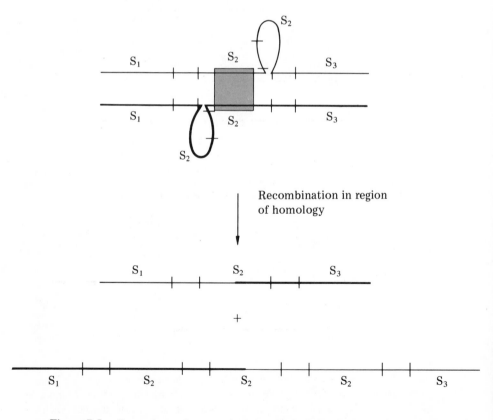

Figure 7-5.   Expansion and contraction of an homologous gene family.

there are increased opportunities for such expansion and contraction. Thus gene duplication increases the possibilities for more gene duplication.

(d)   The evolution of the hemoglobin genes is an example of the refinement of a particular function as a consequence of gene duplication. Hemoglobin is a tetrameric molecule composed of two identical α subunits and two identical β subunits. In humans, the hemoglobin subunits are the products of two unlinked groups of genes, the α-like genes (in humans there may be two α genes) and the β-like genes (β, δ, γ, and possibly ϵ). Generally, it is accepted that vertebrate hemoglobin genes resulted from a series of discrete gene duplications starting with an ancestral gene whose descendents gave rise successively to the myoglobin, α, γ, β, and δ genes (Figure 7-6). Initially, the α chain evolved the ability to interact with itself to form a tetrameric structure under appropriate conditions. Lampreys are still at this evolutionary stage; they have several α-like genes whose products can form tetrameric molecules under conditions of low oxygen tension. In the evolution of human globin genes, one of the α-like genes presumably was translocated to another portion of the genome. The product of

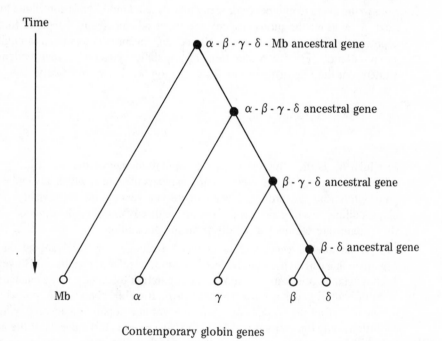

Figure 7-6.   Evolution of the globin genes of man. ● indicates points at which gene duplication occurred. Mb designates myoglobin. The branch lengths are roughly proportional to the time that has passed since the indicated gene duplication.

this translocated gene, while still maintaining its oxygen-storage ability, gradually evolved by nucleotide substitutions to a chain capable of forming stable tetramers with the α chain. In time the β-gene underwent a series of discrete gene duplications to generate the closely linked δ, γ, and (presumably) ϵ genes. This evolutionary process has produced two major improvements: (1) a multisubunit molecule that is capable of modifying its functions through allosteric interactions with physiological effector molecules (e.g., O_2, CO_2, H^+) and (2) several different molecules, produced by different combinations of β-like chains, that have functional properties that are useful to the organism in different developmental stages or physiological states (ϵ-embryonic, γ-fetal, δ- and β-adult).

7-6 Altered genes can become fixed in a population by natural selection or genetic drift

(a) If an altered gene confers an advantage on the individual that carries it, *natural selection* will favor its fixation in the population. Conversely, if an altered gene confers a disadvantage, natural selection will favor its elimination from the population. In the context of natural selection, advantage and disadvantage refer to the average number of progeny that affected individuals contribute to the next generation; the more progeny the more advantageous the gene. Such a reproductive advantage permits an altered gene eventually to spread throughout the population. From considerations of population genetics, it can be demonstrated that the rate of favorable gene fixation (k_f) is approximately

$$k_f = 4N_e s \, m_f \qquad (7\text{-}1)$$

in which N_e is the effective breeding population size of the species, s is the selective advantage of the altered gene, and m_f is the rate at which advantageous gene alterations are produced per gamete per generation. Presumably, most chance alterations of genes confer a selective disadvantage on the organism, and these genes are eliminated rapidly from the population.

(b) If an altered gene confers neither advantage nor disadvantage on the organism that carries it, natural selection cannot select for or against it. However, such a *neutral* gene can be fixed in the population by chance. This chance fixation of neutral genes is termed *genetic drift*. It can be shown from probability theory that of all the variants of a gene present in a population at some point in evolution, only one variant is destined to give rise to all the copies of the gene that will be present in the distant evolutionary future, even in the absence of natural selection. As an illustration, consider a population of equivalent gene variants. Suppose that at each generation, the genes double from N to $2N$, and then half of them are eliminated randomly, so that only the remaining half is passed on to the next generation. Surprisingly, after relatively few generations, very few of the original gene variants remain (Figure 7-7). In time a single gene will become fixed in the population. Since each of these neutral genes has an equal chance, the probability that any one will become fixed is $1/2N_e$, in which $2N_e$ is the number of gametes. If the rate of neutral mutations is m_n per gamete per generation, the number of new neutral alleles introduced each generation should be $2N_e m_n$ and the rate of fixation (k_n) is

$$k_n = (2N_e m_n)/2N_e = m_n \tag{7-2}$$

Thus the net rate of fixation of neutral alleles is independent of population size. However, it can be demonstrated that the time required for fixation of a particular neutral allele is equal to $4N_e$ generations.

(c)   In small populations, neutral or favorable variants often are lost or fixed by nonrandom fluctuations in a very limited pool of chromosomes. Since the fluctuations in gene frequencies are extreme in small populations, the chances for fixation or elimination of an allele are enhanced greatly over the chances in a large population. Fixation of alleles in small populations or reproductively isolated parts of large populations has been termed the *founder principle*. Rapid evolution often is associated with small populations.

(d)   Most mutations, whether or not they are beneficial, are eliminated by chance soon after they arise. Even a slightly favorable mutation has to occur about fifty times, on the average, before it can become established statistically and spread through a population.

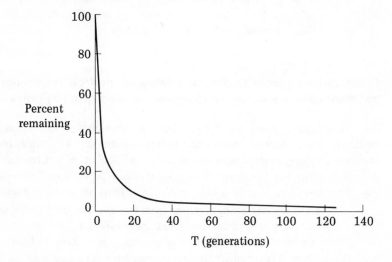

Figure 7-7.   Percent of original gene copies existing at the starting point that remain as a function of chance elimination during successive generations of a population at steady state. This phenomenon is independent of natural selection or rejection. [From R. Doolittle, "Protein evolution," in H. Neurath and R. Hill (Ed.), *The Proteins,* Academic Press, New York, in press.]

(e) The relative evolutionary importance of natural selection and genetic drift is unknown. The debate on this issue constitutes one of the most controversial areas of molecular evolution. The *selectionists* contend that most evolutionary changes are selected, whereas the *neutralists* argue that most changes result from neutral mutations fixed by genetic drift. The range of amino acid replacement rates that must be accounted for by one or both of these processes is extraordinary and is illustrated best by the most conserved and most variable polypeptides studied to date — histone IV and the fibrinopeptides.

(1) Even though plants and animals diverged more than a billion years ago, histone IV (Essential Concept 2-5) from pea and calf differ by only two out of 102 amino acid residues. These two replacements, Ilu/Val and Arg/Lys, are very conservative. Most if not all other amino acid changes apparently disrupt the function of histone IV and consequently have been eliminated by natural selection. If so, the selective pressures must be incredibly precise, since there are 25 other Lys and Arg residues and 14 other Val and Ilu residues that cannot interchange.

(2) The fibrinopeptides, A and B, are removed from fibrinogen by thrombin during the formation of fibrin clots. These peptides prevent the premature polymerization of fibrinogen molecules. The amino acid sequences of fibrinopeptides from species to species are extraordinarily variable (Figure 7-8). Presumably, their function can be accomplished equally well by many different amino acid sequences. If so, a very large number of neutral mutations have been fixed in these peptides by genetic drift.

7-7 Comparative sequence analysis of homologous proteins reveals the ancestral relationships and rates of change of the corresponding genes

(a) *Homologous* genes are derived from a common ancestor by *divergent* evolution. The similarities between homologous genes result primarily from this common ancestry coupled with selection for identical or related functions. Homologous genes can be classified into two types, depending on the mechanism that permitted them to evolve independently of (diverge from) one another.

(1) *Orthologous* genes diverged from one another after speciation, that is, after they were separated into different species and thereby removed from a common gene pool. The ancestry of orthologous genes (e.g., cytochrome c) corresponds with the ancestry of the species that contain them.

(2) *Paralogous* genes diverged from one another after gene duplication, that is, after one gene copy was released from some selection pressures. The ancestry of paralogous genes (e.g., α and β hemoglobin) reflects the history of gene duplication events.

(b)   Homologous genes can be related systematically and quantitatively by construction of *phylogenetic* or *genealogical* trees, as described in the Appendix to this chapter. The purpose of such an analysis is to reconstruct the evolutionary history of homologous genes by determining the minimum number of mutational events that are necessary to relate all contemporary sequences to a single common ancestor (Figures 7-9 and 7-10). The branch lengths of the resulting treelike structures are proportional to the number of substitutions, insertions, and deletions that have been fixed between branch points. The branch points indicate speciation or gene duplication depending on whether orthologous (cytochrome *c*, Figure 7-9) or paralogous (hemoglobins, Figure 7-10) genes are being compared.

Fibrinopeptide A

```
19  18  17  16  15  14  13  12  11  10   9   8   7   6   5   4   3   2   1

                            Lys
                        Val Ala Lys
                Thr Asp Ser Gln Glu
      Asp       Gly Asn Thr Glu Asp                 Ala
Ala   Glu Gly   Pro Glu Ala Thr Gly Gly             Thr Glu     Gly     Val Val
Thr   Gly Asp   Ser Lys Pro Ser Ser Glu Phe Ilu     His Ala Gly Ala Gly Gly         Arg
Glu   Thr       Ala Val Gln Val Asp Ser Leu         Ser Lys         Asp     Gly
                        Val                         Glu
                        Pro
                        Asp
```

Fibrinopeptide B

```
21  20  19  18  17  16  15  14  13  12  11  10   9   8   7   6   5   4   3   2   1

                                            Gly             Leu
            His Tyr                         Lys             Gly
    Ilu     Gly Asp                 Gly     Ala         Ser Thr Val
Ilu Leu Asp Thr         Thr Asp Glu Glu Asp         Asp Val Gly Val     Asp Ala
Glp His Ser Ala Tyr Asp His Glu Gly Asp Glu Arg     Ilu Lys Ala Phe Phe Gly Gly Arg
His Pro Ala Ser Asp Tyr Glp Asp Glu Asn Asn Asp     Thr Glu Val Phe Ala Leu Ser Val
Arg Tyr Pro Ilu         Gly Val Gln     Tyr         Pro     Leu Ala Ilu
    Phe Leu Leu                                     Val     Arg
                                                    Glu     Pro
                                                            His
```

Figure 7-8.   Amino acid residue alternatives found in fibrinopeptides from 43(A) and 41(B) species of mammals. Glp indicates pyrrolidone carboxylic acid. Numbering begins at the invariant C-terminus of these peptides. [Adapted from R. Doolittle, "Protein evolution," in H. Neurath and R. Hill (Ed.), *The Proteins,* Academic Press, New York, in press.]

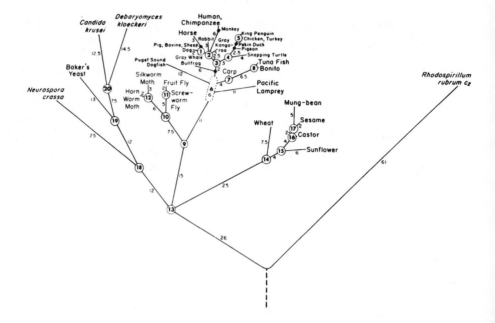

Figure 7-9. Phylogenetic tree for cytochrome c. The numbers of inferred amino acid changes per 100 residues are shown at the branches of the tree. The circled numbers at nodal points indicate ancestral sequences that can be inferred. [From M. O. Dayhoff, C. M. Park, and P. J. McLaughlin, "Building a phylogenetic tree: cytochrome c," in M. O. Dayhoff (Ed.), *Atlas of Protein Sequence and Structure* Vol. 5, National Biomedical Research Foundation, Washington, D.C., 1972.]

(c) Genealogical analysis of many different evolutionarily related proteins suggests a number of conclusions about ancestral relationships and rates of genetic change.

    (1)   Protein families evolve at different but more or less characteristic rates (Table 7-2). Presumably, those genes whose proteins have the most precise structure-function requirements tend to evolve most slowly.

    (2)   Rates of amino acid replacement probably are related more closely to generation time than to absolute time. Amino acid substitution rates depend in part on mutational rates and should be expressed in the same units (substitutions/DNA replication). However, it is difficult to evaluate generation time in terms of DNA replication, since one generation in higher animals often involves 30 to 50 rounds of DNA replication during gametogenesis.

    (3)   Generally, phylogenetic trees that are derived from orthologous protein sequences resemble closely the classical trees derived from fossil records and taxonomic studies (e.g., Figure 7-9). Thus molecular paleontology is consistent with classical paleontology.

    (4)   Primitive ancestral sequences, which are derived from genealogical

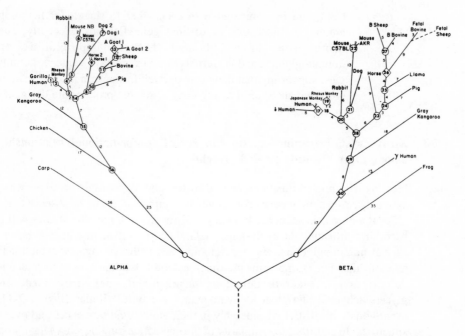

Figure 7-10. Evolutionary tree of hemoglobin α and β chains. The number of inferred amino acid changes per 100 residues are shown at the branches of the tree. Circles indicate species divergence and diamonds indicate gene duplications. The numbers within circles or diamonds indicate ancestral sequences that can be inferred. [From M. O. Dayhoff, L. T. Hunt, P. J. McLaughlin, and D. D. Jones, "Gene duplications in evolution: the globins," in M. O. Dayhoff (Ed.), *Atlas of Protein Sequence and Structure* Vol. 5, National Biomedical Research Foundation, Washington, D.C., 1972.]

Table 7-2. Rates of evolution in different protein families. Unit evolutionary period is the time in millions of years needed to establish a 1% difference between divergent lines. (From R. Dickerson, *J. Mol. Evol.* **1**, 26, 1971.)

Protein Families	Unit evolutionary period
Fibrinopeptides	1.1
Proinsulin (midpiece)	1.9
Ribonuclease (animal)	2.1
Hemoglobin	5.8
Trypsinogen	11
Insulin	10-20
Glyceraldehyde 3-phosphate dehydrogenase	18
Cytochrome c	20
Histone IV	600

trees, can be compared with each other to search for homology among the ancestors of distinct gene families. Recently, such comparisons have indicated that the haptoglobins, the immunoglobins, and the serine proteases, do have sequence homologies, suggesting that they were derived from a common, even more primitive ancestor. These groups of related proteins are designated *superfamilies*.

7-8 X-ray crystallographic studies can reveal conformational relationships among more distantly related proteins

Proteins with related functions, whose corresponding genes diverged very long ago, may retain similar overall or active-site conformations in the absence of significant amino acid sequence homology. X-ray crystallographic studies will be useful in confirming the evolutionary relationships of distantly related proteins, such as immunoglobulins and haptoglobins, and in discerning new evolutionary relationships. For example, the pyridine-nucleotide binding sites of liver alcohol dehydrogenase, lactate dehydrogenase, malate dehydrogenase, and glyceraldehyde-3-phosphate dehydrogenase are quite similar (Figure 7-11), even though they differ considerably in their amino acid sequences and overall structures. In addition, the similarity of the nucleotide binding site of flavodoxin suggests that it also may be related, though more distantly, to the dehydrogenases. Perhaps the conformation of this site is one of the earliest and most universal architectural units of proteins, since the requirement for nucleotide binding must have existed even in the most primitive life forms.

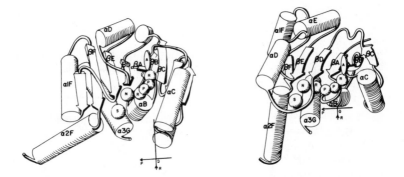

Figure 7-11. Diagrammatic comparison of the coenzyme binding sites of glyceraldehyde-3-phosphate dehydrogenase (left) and lactate dehydrogenase (right). The letters A, P, R, and N indicate adenine, phosphate, ribose, and nicotinamide, respectively. S indicates an essential half cysteine residue. Note that the basic components of tertiary structure for these sites, that is, the helices (α) and the β sheets (β), are conserved. (From M. Buehner *et al.*, *Proc. Natl. Acad. Sci. (U.S.)* **70**, 3052, 1973.)

7-9   Rates of evolution of different DNA classes can be studied by DNA-DNA reassociation

(a)   The separated frequency classes of DNA (satellite, repetitive, and unique) from different species can be compared by DNA-DNA reassociation experiments. The nucleotide sequence divergence of two DNA's can be determined by measuring the difference in melting temperature of native DNA and hybrid duplex DNA after reassociation. Every 1.5% mismatch of nucleotide sequence will produce about a 1°C drop in the melting temperature of hybrid. If. the divergence time of the two species is known, the rate of nucleotide change can be calculated.

(b)   DNA-DNA reassociation studies involving a variety of animals indicate that repetitive and unique DNA classes evolve at about 10^{-10} nucleotides per base pair per gamete division, which is approximately the rate calculated for the fibrinopeptides. In contrast, satellite DNA's may evolve much more rapidly as illustrated by the satellite DNA's in different mouse species. *Mus musculus, Mus caroli,* and *Mus cervicolor* each contain about 10^6 copies of a DNA sequence that is 300 nucleotides long. The sequences of these DNA's are highly divergent from species to species. Either different, though similar, satellite DNA's arose independently in these species, or once identical sequences are diverging very rapidly. Since these species diverged only in the past few million years, either alternative poses difficult questions. Mechanisms for such sudden events (on an evolutionary scale) are unknown. However, since satellite DNA's are common throughout the animal kingdom, such events must be common, if not important, in evolution.

7-10   Certain characteristics of multigene families challenge some current concepts of evolution

(a)   A multigene family is a closely linked cluster of genes that possess sequence homology and perform similar or identical functions. Multigene families have two intriguing evolutionary properties: The size of a family can change rapidly in terms of evolutionary time *(gene expansion and contraction)* and each member of a family in a species evolves so that identical sequences or sequence patterns are maintained in all members *(coincidental evolution).*

(b)   Three general categories of multigene families have been defined, as shown in Table 7-3.

(1)   *Simple-sequence gene families* are found in many animal species. Simple-sequence families can often be isolated as satellites from the bulk of the genomic DNA by buoyant density centrifugation. Their functions are unknown. For example, the three satellite sequences of *Drosophila virilis* are given in Figure 7-12. These heptanucleotide sequences are related to one another by single nucle-

Table 7-3. Categories of multigene families. (Adapted from L. Hood et al., *Ann. Rev. Genetics*, in press, 1975.)

Category	Sequence homology	Number of copies	Gene products	Localization	Example
Satellite	95%-100%	10^3-10^6	None known	Centromere, telomere, other hetero-chromatic regions	Mouse satellite DNA
Multiplicational	99%-100% (identical)	10^1-10^3	RNA, protein	Euchromatin	5S, 18S, and 28S rRNA
Informational	50%-99% (nonidentical)	10^1-10^4	Protein	Euchromatin	Antibodies

Satellite	Sequence
I	$5' \ldots$ ACAAACT $\ldots 3'$
II	$5' \ldots$ ATAAACT $\ldots 3'$
III	$5' \ldots$ ACAAATT $\ldots 3'$

Figure 7-12. The satellite DNA sequences (light strand) from *Drosophila virilis*. (Data from J. Gall *et al.*, *Cold Spring Harbor Symposia on Quantitative Biology* **38,** 417, 1973.)

otide substitutions, suggesting that they diverged from a common ancestor. *Drosophila americana americana* has one major satellite with a heptanucleotide sequence identical to that of satellite I from *Drosophilia virilis*. The presence of identical satellite DNA's in separate species raises questions as to how these satellites arose and how their homogeneity was preserved over a time period adequate for speciation.

(2) *Multiplicational gene families* carry out functions that require many copies of the gene product at some stage in development or in the cell cycle. Two multiplicational gene families, ribosomal and 5S genes, have been well characterized in terms of comparative DNA structure in the amphibian species, *Xenopus laevis* and *Xenopus mulleri*. The general structures of the repeating units in these families are shown in Figure 2-15. The 18S and 28S rRNA genes in both species are identical and tandemly repeated about 450 times. Though the spacers within each species are identical, they differ by more than 1000 out of 7000 nucleotides between species (Figure 7-13). The 5S genes in both species also appear to be identical; however, *X. mulleri* has 9000 repeats whereas *X. laevis* has 24,000. Within each species, the spacers are similar (95%-98% identity), but between the two species they differ in size by a factor of 2.5 and do not cross-hybridize at all.

(3) *Informational gene families* are those whose members differ in sequence so that each can carry out a slightly different function. In contrast, the satellite and multiplicational families have very similar if not identical members. Antibody genes are the only documented example of an informational family.

(c) A variety of mechanisms has been proposed to explain gene expansion and contraction, and coincidental evolution.

(1) Natural selection, while at first a logical possibility, fails to explain either phenomenon. For example, it is difficult to understand how natural selection could select against one altered 28S ribosomal gene in the midst of 449 functioning copies, especially when *Xenopus* can survive perfectly well with half as many ribosomal genes. Furthermore, how could natural selection permit spacers to accumulate mutations rapidly, giving rise to large interspecies differences, yet keep them very homogeneous within a species?

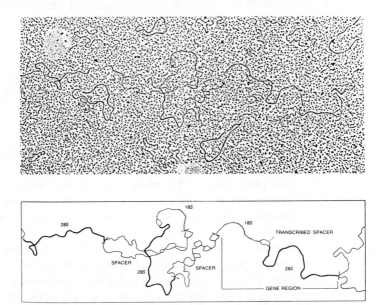

Figure 7-13. A heteroduplex molecule consisting of one strand of ribosomal DNA from *X. laevis* and another from *X. mulleri* (top). Where the nucleotide sequences match, the two strands reassociate; where they do not match, the strands remain separated forming looped regions, as shown in the tracing (bottom). The reassociated and the mismatched regions are the gene and spacer regions, respectively. (Courtesy of Dr. N. Davidson.)

(2) A variety of gene correction mechanisms have been proposed for maintaining identical genes within a family. A typical model, the master-slave model [Essential Concept 2-15(a)], suggests that multiple repeats are tested periodically against a master by base pairing, and deviations in sequences are corrected. Such a correction process would maintain repeat homogeneity and still permit the introduction of single nucleotide substitutions (in the master copy) such as those seen with *D. virilis* satellites. Correction models fail to explain coincidental evolution in nonidentical multigene families such as the antibody genes.

(3) *Unequal crossing-over* can account for rapid expansion and contraction of gene families and coincidental evolution in identical and nonidentical multigene families. Crossing-over in a misaligned region among homologous genes could change rapidly the size of a multigene family (Figure 7-5). Furthermore, if a family of tandem repeats undergoes a sufficient number of such crossover events, then all genes in the family will be descended from a single ancestor (Figure 7-14). This process is termed *crossover fixation*. If the rate of

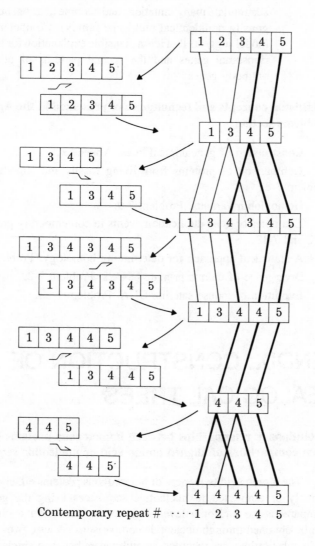

Figure 7-14. Schematic illustration of crossover fixation. Each box represents a repeat, and the number inside each box refers to the ancestor from which the repeat was derived (top line). Lines show the evolutionary descent of repeats. Crossover events are indicated at each level on the left (Adapted from G. Smith, *Cold Spring Harbor Symposia on Quantitative Biology* **38,** 507, 1973.)

crossing-over is frequent, the crossover fixation time will be too short for the repeats to accumulate many mutations and the family of genes will be homogeneous (an identical multigene family). By contrast, if the crossover fixation time is long, the repeats will accumulate many mutations and become heterogeneous but homologous (a nonidentical multigene family). Computer analysis of this model suggests that it is a plausible explanation for the identity of the ribosomal genes and the special patterns of nonidentity in the antibody genes.

7-11   Additional concepts and techniques are presented in the Appendix and the Problems section

(a)   Construction of genealogical trees. Appendix.

(b)   Comparison of proteins from living fossils and "modern" organisms. Problem 7-8.

(c)   Hemoglobin variants. Problem 7-9.

(d)   Examples of gene duplication events in contemporary proteins. Problems 7-12 and 7-13.

(e)   A statistical approach for determining homology. Problem 7-20.

(f)   Divergence of unique primate DNA's. Problem 7-22.

(g)   Evolution of mouse satellite DNA. Problem 7-23.

APPENDIX: CONSTRUCTION OF GENEALOGICAL TREES

A7-1   Evolutionary relationships between homologous proteins can be inferred from comparisons of aligned amino acid or nucleotide sequences

(a)   The amino acid sequences of homologous proteins either can be compared directly or converted to nucleotide sequences using the genetic code and compared at the nucleotide level. More information about evolutionary changes can be obtained through nucleotide comparisons because two- and three-nucleotide substitutions are recorded as multiple rather than single events.

(b)   For valid comparison, sequences must be aligned so as to maximize homology. Homology is maximized by aligning sequences so that the minimum number of genetic events (nucleotide substitutions or sequence gaps) are required to interconvert them. The one-letter amino acid code facilitates such comparisons (Table 7-4). Some examples of proper sequence alignment are shown in Figure 7-15. The placement of gaps reflects the insertion or deletion of one or more

Table 7-4. Amino acid abbreviations and codons. X = any base, Y = either pyrimidine, and Z = either purine.

Three-letter abbreviation	Single-letter abbreviation	mRNA codons
Ala	A	GCX
Arg	R	CGX, AGZ
Asn	N	AAY
Asp	D	GAY
Cys	C	UGY
Glu	E	GAZ
Gln	Q	CAZ
Gly	G	GGX
His	H	CAY
Ilu	I	AUY, AUA
Leu	L	CUX, UUZ
Lys	K	AAZ
Met	M	AUG
Phe	F	UUY
Pro	P	CCY
Ser	S	UCX, AGY
Thr	T	ACX
Trp	W	UGG
Tyr	Y	UAY
Val	V	GUX

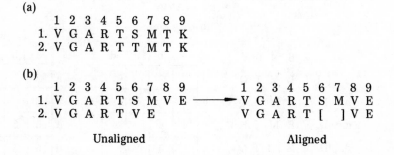

(a)
```
      1 2 3 4 5 6 7 8 9
   1. V G A R T S M T K
   2. V G A R T T M T K
```

(b)
```
      1 2 3 4 5 6 7 8 9          1 2 3 4 5 6 7 8 9
   1. V G A R T S M V E  ——→   V G A R T S M V E
   2. V G A R T V E              V G A R T [   ] V E

         Unaligned                    Aligned
```

Figure 7-15. Examples of sequence alignment. (a) Two polypeptides that are related by a single amino acid interchange. (b) Two polypeptides that are related by a gap.

codons in homologous genes. Gap placement is difficult in distantly related proteins, and must be determined by careful statistical criteria.

A7-2 A difference matrix converts sequence alignments into quantitative relationships

A difference matrix tabulates the number of substitutions, insertions, and deletions that are required to relate all possible pairwise comparisons in a set of homologous proteins. Each gap is counted as one genetic event. An example of the construction of a difference matrix is shown in Figure 7-16. Sequence *a* differs from Sequences *b, c,* and *d* by 2, 4, and 4 amino acid residues, respectively, at the protein level, and by 2, 5, and 5 genetic events, respectively, at the nucleic acid level. Sequence analysis at the nucleic acid level is carried out using the minimum number of substitutions required. For example, if one codon can be converted to another by a 1 or 2 base change, the minimum difference always is recorded.

A7-3 Genealogical trees depict minimum evolutionary pathways

(a)   Genealogical analysis attempts to determine the minimum number of genetic events that are required to generate a homologous set of proteins from a single ancestral sequence. This minimum evolutionary path is presented as a tree in which contemporary sequences (e.g., *a, b, c,* and *d* in Figure 7-16) are represented as terminal twigs (Figure 7-17). Starting from two adjacent twigs (e.g., *a* and *b*), a series of nodal sequences *(e* and *f)* can be derived until the most primitive sequence, the ancestral sequence *(g),* is reached. Presumably, *g* is the sequence from which each of the contemporary sequences originally was derived. As indicated in Essential Concept 7-7(b), branch points reflect speciation or

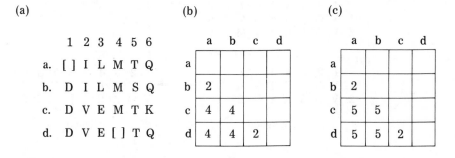

Figure 7-16. Construction of a difference matrix. (a) Aligned amino acid sequences. (b) Amino acid difference matrix. (c) Nucleotide difference matrix.

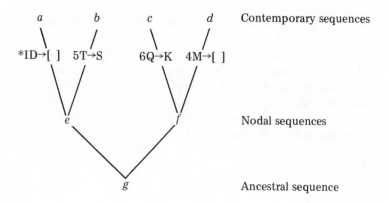

Figure 7-17.   A genealogical tree of Sequences *a, b, c,* and *d* in Figure 7-16. The asterisk indicates that at Position 1 in Protein *a,* D was deleted.

gene duplication and branch lengths are proportional to the number of substitutions, insertions, and deletions.

(b)   The inspection method of constructing genealogical trees starts from an amino acid difference matrix. Sequences that are related most closely are placed on adjacent branches, as shown in Figure 7-17 for Sequences *a, b, c,* and *d* of Figure 7-16. A preliminary nodal sequence then can be determined for *e* by a residue-to-residue comparison of the two sequences, *a* and *b,* that branch from this node (Figure 7-18). Where both proteins are identical, the nodal sequence must be the same. Where *a* and *b* differ, the nodal sequence could be either one. The same procedure can be repeated for nodal sequence *f.* The ancestral sequence, *g,* can be determined in a similar way using the preliminary nodal sequences, *e* and *f* (Figure 7-18). Certain ambiguities in Sequence *g* cannot be resolved with this information. However, the unambiguous residues in *g* can be used to resolve all the ambiguities in *e* and *f* (Figure 7-18). These final nodal sequences permit specific mutation events to be assigned to each branch as indicated in Figure 7-17.

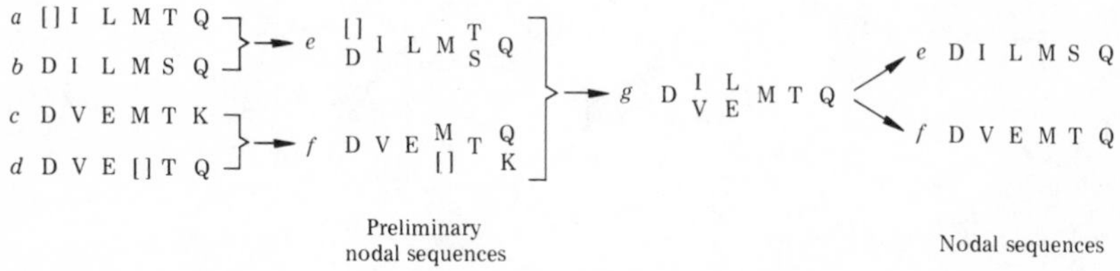

Figure 7-18.   Example of the inspection method of construction genealogical trees.

(c)   The Fitch-Margoliash procedure, which is a quantitative method for generating genealogical trees, starts from a nucleotide difference matrix. The basic approach is illustrated in Figure 7-19, using Sequences *a, b, c,* and *d* from Figure 7-16. Initially, the two sequences with the fewest mutational differences are placed on Branches I and II; all other sequences are placed on Branch III. The length of each branch then is calculated in terms of the mutational distances derived from the nucleotide difference matrix [Figure 7-19(a)]. In subsequent steps the sequence on Branch II is combined with the sequence on Branch I, the

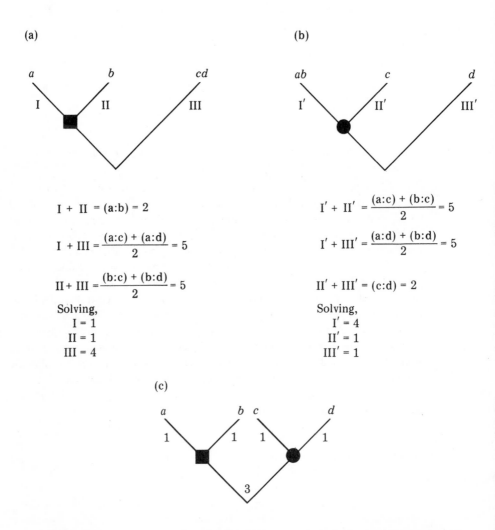

(a)

$$I + II = (a:b) = 2$$

$$I + III = \frac{(a:c) + (a:d)}{2} = 5$$

$$II + III = \frac{(b:c) + (b:d)}{2} = 5$$

Solving,
  I = 1
  II = 1
  III = 4

(b)

$$I' + II' = \frac{(a:c) + (b:c)}{2} = 5$$

$$I' + III' = \frac{(a:d) + (b:d)}{2} = 5$$

$$II' + III' = (c:d) = 2$$

Solving,
  I' = 4
  II' = 1
  III' = 1

(c)

Figure 7-19.   Example of the Fitch-Margoliash method of constructing genealogical trees. (a) Step one. (b) Step two. (c) Final genealogical tree for Sequences *a, b, c,* and *d.* The black circle and square relate the nodes in (a) and (b) to those in (c).

next most closely related sequence is placed on Branch II, all other sequences are placed on Branch III, and a new set of branch lengths are calculated [Figure 7-19(b)]. The branch lengths from a series of such calculations then can be arranged into a tree as shown in Figure 7-19(c).

This type of genealogical tree is a graphical representation of the order in which ever more-distantly related sequences are added. For a small number of sequences, the appropriate order of addition often can be determined by visual inspection. As the number of sequences increases, the order in which they should be added to the genealogical tree becomes more difficult to determine. Because of the averaging process that is employed in the Fitch-Margoliash procedure, addition of sequences to the tree in different orders will generate trees with distinct topologies. A computer can be used to generate and examine many such trees to find which tree minimizes the number of genetic events.

A7-4   Divergence rates of protein families can be estimated from genealogical trees

(a)   Since the probability of multiple substitutions in a single codon increases as genes diverge, a statistical correction often is applied before divergence rates are calculated. If a random distribution of substitutions is assumed, then

$$\frac{m}{100} = -\ln(1 - \frac{n}{100})$$ (A7-1)

in which n represents the observed number of amino acid differences per 100 residues and m represents the corrected number of amino acid differences per 100 residues.

(b)   Given the divergence time of two species and the number of amino acid (or nucleotide) differences between two homologous proteins, a rate of evolutionary change can be calculated. This rate, designated the *unit evolutionary period* (UEP), is defined as the time required for two proteins to diverge by 1% of their amino acid sequence. Thus

$$\text{UEP} = \frac{\text{divergence time}}{m}$$ (A7-2)

in which the divergence time can be absolute time or generation time. This calculation is a useful first approximation for estimating rates of species evolution but suffers from simplifying assumptions (e.g., that generation times and evolutionary rates of the species compared are similar).

REFERENCES

Where to begin

R. Dickerson, "The structure and history of an ancient protein," *Scientific American* (April, 1972)

R. Doolittle, "Protein evolution," in H. Neurath and R. Hill (Ed.), *The Proteins* (Academic Press, New York, in press, 1976)

R. Stroud, "The protein cutting enzymes," *Scientific American* (July, 1974)

E. Zuckerkandl, "The evolution of hemoglobin," *Scientific American* (May, 1965)

General

M. Dayhoff, *Atlas of Protein Sequence and Structure* (National Biomedical Research Foundation, Washington D. C., 1972)

R. Dickerson and I. Geis, *The Structure and Action of Proteins* (W. A. Benjamin, Menlo Park, Calif., 1969)

T. Dobzhansky, *Genetics of the Evolutionary Process* (Columbia University Press, New York, 1970)

R. Lewontin, *The Genetic Basis of Evolutionary Change* (Columbia University Press, New York, 1974)

S. Ohno, *Evolution by Gene Duplication* (Springer-Verlag, Berlin and New York, 1970)

E. Smith, "Evolution of enzymes," in P. Boyer (Ed.), *The Enzymes* (Academic Press, New York, 1970)

The selectionists and neutralists

B. Clarke, "Darwinian evolution of proteins," *Science* **168,** 1009 (1970)

J. Crow, "The dilemma of nearly neutral mutations: how important are they for evolution and human welfare?," *J. Heredity* **63,** 306 (1972)

G. Johnson, "Enzyme polymorphism and metabolism," *Science* **184,** 28 (1974)

M. Kimura and T. Ohta, "Protein polymorphism as a phase of molecular evolution," *Nature* **229,** 467 (1971)

J. King and T. Jukes, "Non-Darwinian evolution," *Science* **164,** 788 (1969)

Evolution of DNA

D. Brown and K. Sugimoto, "The structure and evolution of ribosomal and 5S DNA in *Xenopus laevis* and *Xenopus mulleri*," *Cold Spring Harbor Symposia on Quantitative Biology* **38,** 501 (1973)

J. Gall, E. Cohen, and D. Atherton, "The satellite DNA's of *Drosophila virilis*," *Cold Spring Harbor Symposia on Quantitative Biology* **38,** 417 (1973)

D. Kohne, "Evolution of higher-organism DNA," *Quart. Rev. Biophysics* **33,** 327 (1970)

C. Laird, B. McConaughy, and B. McCarthy, "Rate of fixation of nucleotide substitutions in evolution," *Nature* **224,** 149 (1969)

H. Smith (Ed.), *Evolution of Genetic Systems* (Gordon and Breach, New York, 1972)

Evolution of multigene families

G. Edelman and J. Gally, "Arrangement and evolution of eukaryotic genes," in F. O. Schmitt (Ed.), *The Neurosciences Second Study Program* (Rockefeller University Press, New York, 1970)

L. Hood, J. Campbell, and S. Elgin, "The organization, expression and evolution of antibodies and other multigene families," *Ann. Rev. Genetics* (in press, 1975)

G. Smith, "Unequal crossover and the evolution of multigene families," *Cold Spring Harbor Symposia on Quantitative Biology* **38,** 507 (1973)

C. Thomas, "The rolling helix: a model for the eukaryotic gene," *Cold Spring Harbor Symposia on Quantitative Biology* **38,** 347 (1973)

C. Thomas, Jr., "The theory of the master gene," in F. O. Schmitt (Ed.), *The Neurosciences Second Study Program* (Rockefeller University Press, New York, 1970)

Sequence analysis and genealogical trees

L. L. Cavalli-Sforza, "Some current problems of human population genetics," *Amer. J. Hum. Genet.* **25,** 82 (1973)

M. Dayhoff, "Computer analysis of protein evolution, *Scientific American* (July, 1969)

R. Dickerson, "The structure of cytochrome *c* and the rates of molecular evolution," *J. Mol. Evol.* **1,** 26 (1971)

W. Fitch and E. Margoliash, "Construction of phylogenetic trees," *Science* **155,** 279 (1967)

W. Fitch and E. Markowitz, "An improved method for determining codon variability in a gene and its application to the rate of fixation of mutations in evolution," *Biochem. Gen.* **4,** 579 (1970)

S. Needleman and C. Wunsch, "A general method applicable to the search for similarities in the amino acid sequences of two proteins," *J. Mol. Biol.* **48,** 443 (1970)

PROBLEMS

7-1 Answer the following with true or false. If false explain why.

(a) Natural selection limits the mutations that can occur in the genome.

(b) The most common way for a gene to change during evolution is by single-base substitutions.

(c) Homologous proteins have similar amino acid sequences, but have different evolutionary origins.

(d) Mutations accumulate at approximately equal rates in unique and middle-repetitive DNA's.

(e) Different families of proteins can evolve at different rates.

(f) Genes that evolved by discrete gene duplication should contain internal homology.

(g) Genealogical trees can be constructed from nucleic acid or protein sequence information.

(h) Generally, radical amino acid substitutions permit a protein to function more effectively.

(i) Sequence gaps are less frequent evolutionary events than single-base changes.

(j) The number of residues by which two homologous proteins differ is an overestimate of the number of genetic events that have occurred during their evolution.

(k) The amino acid replacement Phe/Trp is a frequent conservative substitution in proteins.

(l) Hybrid gene formation results from a crossover event between non-identical genes.

(m) The net rate of fixation of neutral mutations in a population is $4N_e$ generations.

(n) In general, genealogical trees constructed from comparisons of homologous proteins agree with more classical phylogenetic trees.

(o) The structure of the genetic code favors conservative amino acid substitutions.

(p) Duplicated genes rapidly can increase or decrease in number by recombination in misaligned regions that result from nonhomologous pairing.

(q) Rapid evolution often is associated with small populations.

(r) Natural selection tends to eliminate neutral mutations, though at a very slow rate.

(s) The majority of favorable mutations are eliminated by chance during the first few generations after they arise.

7-2 (a) _____ proteins have a common evolutionary origin.

(b) _____ leads to a multiplication of the genome size.

(c) _____ and _____ are the two mechanisms by which altered genes can become fixed in a population.

(d) The two kinds of genetic alterations that appear to be most important in evolution are _____ and _____.

(e) Small scale gene duplications probably are produced by a process known as _____.

(f)   A _____ mutation produces a codon that codes for the same amino acid as the nonmutated codon.

(g)   If both crossover points are outside the structural gene, a _____ gene duplication will be produced in one of the resulting chromosomes.

(h)   One mutant of *E. coli* makes an altered _____ that increases the mutation rate by a factor of 100 to 1000.

(i)   _____ amino acid interchanges involve amino acids with similar side chains.

(j)   A _____ tabulates amino acid or nucleotide differences that exist in all pairwise combinations of a group of homologous proteins.

(k)   _____ and the _____ are, respectively, the most conserved and most variable polypeptides studied to date.

(l)   _____ proteins diverged from one another after speciation, whereas _____ proteins diverged from one another after gene duplication.

(m)   A _____ is an attempt to retrace the evolutionary events that produced a set of contemporary homologous proteins from a single ancestral protein.

(n)   _____ is a term that designates the phenomenon whereby members of a multigene family evolve together.

(o)   The rapid changes in the size of a multigene family are designated

_____ .

(p)   _____ studies can reveal conformational relationships among more distantly related proteins.

(q)   The three kinds of multigene families are _____ , _____ , and _____ .

7-3   (a)   Determine by inspection how many sets of homologous proteins are given in the following list. (All sequences in this chapter are given in the standard orientation, N terminus or 5′ terminus at the left end.)

(1)   G D V E K G K K I F I M K C S
(2)   V L S P A D K T N V K A A W G
(3)   V H L T P E E K S A V T A L W
(4)   A D S G E G D F L A E G G G V
(5)   G D V E K G K K I F V Q K C A
(6)   V H L T P E E K N A V T T L W
(7)   V L S G E D K S N I K A A W G
(8)   V H L S S E E K S A V T A L W

(b)   Are there any less obvious homology relationships between individual sets?

7-4   Construct a nucleotide difference matrix for the following nucleotide sequences.

(1) A T G C G
(2) A C G T A
(3) G T A C T
(4) T G A C C

7-5 Derive the most probable ancestral sequences for each of the following sets of proteins.

    I. (1) P A P F E Q G S A K K G A T L
         (2) P D P Y E Q G S V R K A G T L
         (3) P A H C E R G S A K R G A T L

    II. (1) Y D L T Q P P S V S V S P G Q
        (2) P D L T Q N K S T S V S P N Q
        (3) Y D L H Q P P Q Y S Y S P G N

    III. (1) V L S E G E W Q L V L H V W A
        (2) I L S K G E Y Q L A L H V W A
        (3) V L S E G N W Q L V L H A Y A

7-6 (a) Align the following sequences to maximize their homology.
      (1) C S N L S T C V L S A Y W R N L N N F H
      (2) C S T C V L S A Y W R N L N N F H
      (3) Q C Y L S A Y W F N A N Q F H
      (4) C S N L S T W R N I N Q F H
      (5) S N L S T C L V S A Y W F H
      (6) F T D V T S C L L F W K Q V Y R

(b) Construct amino acid and nucleotide difference matrices for these sequences. (Each gap counts as a twenty-first amino acid or fifth nucleotide.)

(c) Are there any significant discrepancies in the two methods of comparing these sequences?

(d) Which comparison is a more accurate estimate of the evolutionary differences among these sequences? Why?

7-7 (a) Consider a phylogenetic tree based on paleontological data for animals I, II, III, and IV (Figure 7-20). If $A = B + C = B + D + E = B + D + F$, the tree is said to be symmetrical with regard to mutations. Suggest three reasons why such a genealogical tree might not be symmetrical with regard to mutational occurrences.

(b) If $a = b + c = b + d + e = b + d + f$, the tree is said to be symmetrical with regard to fixations. Suggest three reasons other than those listed in (a) why such a tree might not be symmetrical with regard to fixations.

(c) In general, would you expect branch lengths to be more nearly symmetrical if all fixed mutations were neutral or if all were advantageous?

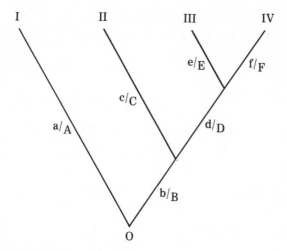

I II III IV

e/E f/F

c/C

a/A d/D

b/B

O

Figure 7-20. Hypothetical genealogical tree (Problem 7-7). Upper and lower case letters designate, respectively, the number of mutations and number of fixations that occurred along different branches. Roman numerals indicate four contemporary species. [From R. Doolittle, "Protein evolution," in H. Neurath and R. Hill (Ed.), *The Proteins*, Academic Press, New York, in press.]

7-8   As a Narcotics Agency Research Consultant, you have become interested in an LSD-binding protein that can readily be isolated by affinity chromatography from the serum of pigs. This protein is a small molecule composed of two polypeptide chains α and β, whose sequences are given in Figure 7-21a.

(a)   Outline a plausible scheme for the evolution of the α and β genes. Be as precise and quantitative as you can in the documentation of your evolutionary scheme.

(b)   Because the LSD-binding serum protein is unusually small for a blood protein, you decide to attempt an isolation directly from the serum based on size fractionation using gel filtration. You succeed in isolating a molecule that upon reduction and alkylation breaks into polypeptides with molecular weights similar to those of α and β. However, fiftyfold more protein is isolated than expected based on the LSD-binding activity of the pig serum. Sequence analysis of the putative α chain reveals a heterogeneity for the N-terminal 10 residues which is given in Figure 7-21b. Suggest a reasonable genetic explanation for these results.

a)

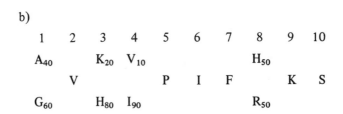

b)

Figure 7-21. (a) The amino acid sequences of the α and β chains of the LSD-binding protein from pig serum. (b) The N-terminal 10 residues of the α-like chains separated from serum by gel filtration (see text). Subscripts indicate the approximate percentages of each residue at positions that exhibit heterogeneity (Problem 7-8).

7-9   A large number of hemoglobin variants have been detected either by electrophoretic screening of normal people or by examination of people with blood disorders. Presumably, these amino acid replacements have been observed before the forces of natural selection have had a chance to operate. As of 1972, 214 amino acid replacements involving 144 unique substitutions had been detected (Table 7-5).

(a)   Do these observations support the contention that variant proteins arise predominantly by single-base substitution?

(b)   How do these amino acid replacements differ from those found in evolutionarily related proteins (Figure 7-3)?

(c)   How would you explain the differences in Part (b)?

7-10   The amino acid sequences of several non-catalytic proteins are comprised of repeating units. For example collagen, the protein of connective tissue is composed primarily of the repeating unit $(Gly-X-Pro)_n$, where X is any amino acid,

Table 7-5. Tabulation of amino acid replacements observed in 214 different abnormal human hemoglobins (Problem 7-9). [From R. Doolittle, "Protein evolution," in H. Neurath and R. Hill (Ed.), *The Proteins,* Academic Press, in press.]

Replacement	Number	Replacement	Number
Lys/Glu	28	Ala/Asp	5
Gly/Asp	26	His/Arg	5
Lys/Asn	19	Leu/Arg	5
Tyr/His	17	Leu/Pro	5
Gln/Glu	14	Gln/Arg	4
His/Asp	13	Gly/Glu	4
Ala/Glu	10	Tyr/Asp	3
Asn/Asp	10	Ser/Arg	3
Arg/Gly	8	Gln/His	2
Val/Glu	6	Gln/Lys	2
Thr/Lys	5	Phe/Ser	2
Pro/Arg	5	Others	13
		(one each)	
			214

and fibroin, the structural protein of silk, has the general structure (Gly-Ser-Gly-Ala-Gly-Ala)$_n$. Keratin, the structural protein of wool, hair, and horn contains sequences of the form (Cys-Cys-Gly-Ser-Pro) $_n$. The "antifreeze protein" that some arctic fish produce as a means of lowering the freezing temperature of their blood is composed predominantly of (Ala-Thr-Thr)$_n$ sequences. Offer an explanation as to how these proteins might have evolved.

7-11 (a) How many possible nucleotide sequences for the α-hemoglobin gene differ from the wild-type sequence by a single nucleotide? α hemoglobin is 141 residues long.

(b) Assuming that all base substitutions are equally likely, calculate the relative probabilities of a reverse mutation (one that converts the gene back to the wild-type nucleotide sequence) and a second forward mutation.

(c) In view of these calculations what can you conclude about the probability of parallel evolution in homologous genes?

7-12   The genes for the β and δ human hemoglobin chains are linked closely and presumably resulted from a gene duplication. Two hemoglobin variants that resemble β and δ hemoglobins have been found. Their amino acid differences and occurrences in affected individuals are shown in Table 7-6.

(a)   Suggest a plausible mechanism for how these two variants arose.

(b)   Deduce the order of the genes for β and δ hemoglobin on the chromosomes of normal individuals.

Table 7-6.   Two normal and two variant human hemoglobins (Problem 7-12). All the amino acid differences that exist between the chains are listed. + and − refer, respectively, to presence and absence in individual humans. (From M. Dayhoff, *Atlas of Protein Sequence and Structure*, National Biomedical Research Foundation, Washington, D.C., 1972.)

| | Normal | | Variant | |
Position	β	δ	I	II
9	S	T	T	S
12	T	N	N	T
22	E	A	A	E
50	T	S	S	T
86	A	S	S	A
87	T	Q	Q	T
116	H	R	H	R
117	H	N	H	N
125	P	Q	P	Q
126	V	M	V	M

Presence in individuals				
Normal	+	+	−	−
Abnormal I	−	−	+	−
Abnormal II	+	+	−	+

7-13   Human haptoglobins are heme-carrying proteins composed of two α and two β chains. In humans there are three allelic forms of the α chain: α^{1F}, α^{1S}, and α^2. The amino acid sequences of these proteins are given in Figure 7-22.

(a)   By what mechanism did the α^2 gene most likely arise?

(b)   In the human population there also exist α^2 chains that have K at positions 54 and 113 and that have E at positions 54 and 113. Suggest a likely mechanism by which these two α^2 chains could have been produced from the genes for the alleles described in Figure 7-22.

(c)   Which altered α chain arose first: the one-base substitution allele (either α^{1F} or α^{1S}, depending on which one was the original gene) or the α^2 allele?

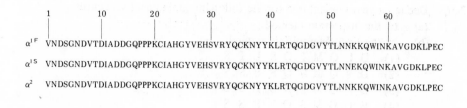

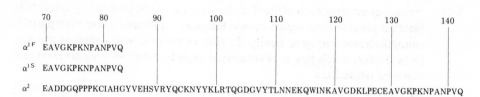

Figure 7-22. The amino acid sequences of the α^{1F}, α^{1S}, and α^2 haptoglobin chains (Problem 7-13). [Adapted from M. O. Dayhoff, *Atlas of Protein Sequence and Structure*, National Biomedical Research Foundation, Washington, D.C. 1972.)

7-14　It is difficult to measure absolute mutation rates. Because most assay systems depend on gross observable defects in protein function or structure, many mutations are not detected. To estimate the difference between measured and absolute mutation rates, you isolate 65 single-base-substitution mutants from the histidine C locus of *Salmonella* (the assay is for protein function) and find that twenty-two of these mutants are base substitutions that result in chain termination codons. Make the simplifying assumptions that all 61 sense codons are equally represented in this gene, that all base changes are equally likely, and that all mutations leading to chain termination have been detected.

(a)　Approximately how many amino acid substitutions occurred in the population sample you examined?

(b)　What percent of the total number of amino acid substitutions did you detect?

(c)　Why did you not find the undetected mutants?

(d)　Approximately how many silent mutations occurred?

7-15　Deduce a genealogical tree for the following nucleotide sequences by the inspection method.

(1)　C A T C A G
(2)　G T T C A G
(3)　G A G T A G
(4)　G A G C A T

7-16   Deduce a genealogical tree for the following amino acid sequences
    (a)   by the inspection method.
    (b)   by the Fitch-Margoliash method.
        (1)   D I Q M T Q S L T S
        (2)   D V Q M S Q S P S S
        (3)   D I Q M T N S L S S
        (4)   E I Q M S Q S P S S

7-17   Histone genes from each of the five classes appear to be encoded by 100 to 400 identical genes in most organisms and therefore constitute another example of a multiplicational multigene family. In view of the multigene nature of the histones, explain briefly how the evolutionary aspects of the following observations might be rationalized.
    (a)   Histones III and IV from two distantly related sea urchins are very similar, if not identical, in amino acid sequence, yet their mRNA's cross hybridize very poorly.
    (b)   Histone IV from pea and cow, organisms that diverged more than a billion years ago, differ by 2 of 102 residues.
    (c)   Two electrophoretically different variants of one histone have been observed in strains of maize.

7-18   (a)   From the divergence times and the average observed number of substitutions per 100 residues (n) in Table 7-7, calculate the unit evolutionary period (UEP) for each pair of proteins and the average UEP of cytochrome c and globin.
    (b)   Why are the average UEP's of cytochrome c and globin different?

Table 7-7.   Divergence times and observed mutational differences/100 residues for various related proteins (Problem 7-18). (From R. Dickerson, *J. Mol. Evol.* **1,** 26, 1971.)

Comparison	n	Divergence (10^6 years)
Cytochrome c		
Birds/reptiles	12.7	240
Mammals/reptiles	14.2	300
Fish/other vertebrates	18.0	400
Vertebrates/insects	24.6	600
Globin		
Between β chains of nonprimates	22.0	90
Mammalian α/carp α	49.8	400
All α/all β, γ, δ	58.1	500

7-19 α-lactalbumin, one component of a lactose synthesis system, evolved from an ancestor common to the lysozyme gene when milk-producing animals developed about 100-150 million years ago. There is significant homology between these proteins. The UEP for lysozyme (from human and chicken sequences) is 5.3×10^6 years. Using this UEP, one can calculate that α-lactalbumin and human lysozyme should have diverged 3.9×10^8 years ago.

(a)   How do you explain the discrepancy between the morphologic data (divergence 1.0×10^8 to 1.5×10^8 years ago) and the molecular data (divergence 3.9×10^8 years ago)?

(b)   A divergence time of 9.4×10^8 years for myoglobin and hemoglobin can be calculated from the UEP for myoglobin. Is this divergence time likely to underestimate or overestimate the real divergence time? Why?

(c)   Is it valid to compare the α, β, γ, and δ hemoglobin chains in calculating the UEP of the globins?

7-20   Homology among more distantly related proteins often is not immediately apparent. In addition, there is always some homology among unrelated proteins. Consequently, statistical approaches have been devised to determine whether the observed homology between proteins most likely exists because of common ancestry or chance. One such approach calculates an average mutational value (AMV) as an indicator. AMV is equal to the minimum number of nucleotide changes required to convert one sequence into another divided by the number of codons compared.

$$AMV = \frac{number\ of\ substitutions}{number\ of\ codons}$$

Analysis of an appropriately weighted 20×20 matrix of amino acids shows that the average number of mutations required to convert one randomly selected amino acid into another is 1.45. The probability that two randomly chosen sequences will have an AMV of 1.13 or less is about 1%.

Using the criterion that an AMV of less than 1.13 indicates probable homology, decide whether the following aligned sequences are homologous.

(a)   (1)   I A H G Y V E H S V R Y Q C K N Y
       (2)   I S S L Q P E D I A T Y Y C Q Q F

(b)   (1)   K V F G R C E L A A M K
       (2)   K E T A A A K F E R Q H

(c)   (1)   E V Q L V E S G G G L I Q P G G S L R L S C
       (2)   Q S A L T Q P P S A S G S P G Q S V T I S C

7-21   The following amino acid sequences were determined for the same protein in eight species:

```
      1       5          10        15         20        25        30          35
S1:     A P V Q D R G E F L A E G G G V R A I D Y D E G E D D R V K V T V
S2:   A D D S D P V G G E F L A E G G G V R Y L D Y D E V D D N R A K L T L
S3:   T D A D A D K G E F L A E G G G V R A T D Y D E E E D D R V K V F L
S4:   T D P D A E E G E F L A E G G G V R A T D Y D E E E D D R V K V F L
S5:     G S D P V G G E F L A E G G G V R Y T D Y D E G D D N R A K L W L
S6:   T D P D A D K G E F L A E G G G V R A T D Y D E E E D D R V K V F L
S7:   T D P D A D E G E F L A E G G G V R A T D Y D E E E D D R R V V F L
S8:   A D D S D P V G G E F L A E G G G V R Y L D Y D E V D D N R A K L T L
```

Construct a protein genealogical tree for these sequences by the inspection method. Indicate all nodal sequences and the amino acid changes that have occurred on each branch.

7-22 A friend has given you a sample of radioactively labeled green monkey unique DNA so you can study the evolution of unique DNA by DNA-DNA reassociation experiments. You mix, denature, and renature labeled green monkey DNA with a large excess of unlabeled green monkey, man, or capuchin DNA to ensure that the labeled DNA forms heteroduplexes rather than homoduplexes. The melting curves for the reassociated DNA's are shown in Figure 7-23.

(a) Calculate the percent nucleotide divergence between green monkey and man. Between green monkey and capuchin. Assume that a 1°C decrease in T_m corresponds to a 1.5% mismatch in base pairing.

(b) The divergence times for green monkey and man and for green monkey and capuchin are $45. \times 10^6$ and 65×10^6 years, respectively. Calculate a UEP for this class of DNA for each pair of primates.

(c) Similar experiments for rat-mouse unique DNA suggest a UEP of 0.25×10^6 years. How might you account for the large difference in UEP values?

7-23 The structure of mouse satellite DNA, which is 10% of the nuclear DNA, should provide some clues about the evolutionary events responsible for creation of the satellite. A partial digestion with Eco RII, a restriction endonuclease that cleaves $\frac{CCAGG}{GGTCC}$ sequences of DNA, breaks mouse satellite DNA into fragments that form a series of major and minor bands when electrophoresed under conditions that separate polynucleotides according to size (Figure 7-24).

(a) The fifth major band from the partial digest was digested again with Eco RII and re-electrophoresed (Figure 7-25). Explain the band patterns in Figures 7-24 and 7-25.

(b) The sixteenth major band in the partial digest has a mobility close to that of the smallest fragments (3840 nucleotide pairs) produced from bacteriophage λ DNA by Eco RI, another restriction endonuclease. What is the molecular weight of the major monomeric unit of the mouse satellite DNA cleaved with Eco RII?

(c) When isolated monomers, dimers, and trimers are denatured, reas-

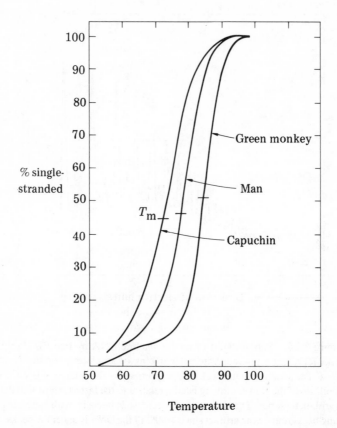

Figure 7-23. Melting profiles of heteroduplex DNA molecules (Problem 7-22). Heteroduplexes were between ^{14}C-labeled green monkey DNA and unlabeled green monkey, man, and capuchin as indicated. T_m indicates the point at which 50% of the heteroduplexes remain associated. ΔT_m for green monkey-man is 7.0°C. and ΔT_m for green monkey-chapuchin is 12.3°C. (Adapted from D. Kohne, *Quart. Rev. Biophys.* **33,** 327, 1970.)

sociated, and remelted, the observed T_m's are about 5°C lower than those observed with the native fragments. How much internal divergence has occurred among the repeated sequences in the satellite? (Assume 1°C decrease in T_m corresponds to 1.5% mismatch of base pairs.)

(d)   Radioactively labeled monomers, dimers, and trimers were isolated from the major peaks. These fractions were disassociated, reassociated, and analyzed by gel electrophoresis. The fraction of monomers that reassociated as linear monomers was 56%; of dimers that reassociated as linear dimers was 25%; and of trimers that reassociated as linear trimers was 17%. In each case the remainder of

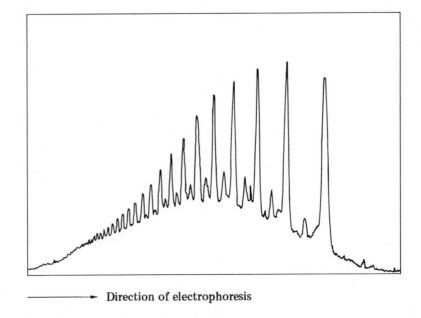

Direction of electrophoresis

Figure 7-24. Partial digest of mouse satellite DNA by Eco RII (Problem 7-23). The peaks represent a microdensitometer tracing of an electrophoretically separated partial digest. The areas under the peaks are proportional to the amount of DNA in each peak. The mobilities of the slower moving bands, relative to the fastest, show that their lengths form an arithmetic series. The slower bands decrease in intensity with increasing digestion. The rising background is an artifact since >90% of the DNA is under the peaks. (Adapted from E. M. Southern, *J. Mol. Biol.*, in press, 1975.)

the DNA reassociated as circles or larger polymers. From these observations calculate the size of the repeat unit for reassociation.

(e)   The minor bands suggest that intermediate forms (or 1/2-mers) are produced by digestion with the restriction enzyme. Given the observations in Part (d) suggest a mechanism for the production of 1/2-mers.

(f)   Partial nucleotide sequence analysis suggests that the mouse satellite has a periodicity of 9 to 18 nucleotide pairs and a longer repeating sequence containing four of these short units. Thus, there are a total of four levels of periodicity in mouse satellite DNA. Assume 15 nucleotide pairs is the smallest periodic unit. Draw a plausible scheme for the evolution of the mouse satellite.

7-24   From the nucleotide difference matrix in Figure 7-26, construct a genealogical tree for globins, using the Fitch-Margoliash method.

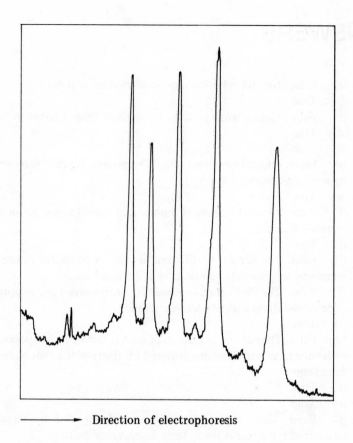

Direction of electrophoresis

Figure 7-25. Redigestion with Eco RII of the fifth major band from the partial digest (Problem 7-23). (Adapted from E. M. Southern, *J. Mol. Biol.*, in press, 1975.)

	α	β	γ	δ	M
α					
β	105				
γ	115	52			
δ	108	12	50		
M	157	167	161	164	

Figure 7-26. Nucleotide difference matrix for globins (Problem 7-24). α, β, γ, and δ refer to different hemoglobins and M refers to myoglobin. This matrix is derived from gap corrected alignments of the globins. (W. Fitch, personal communication.)

ANSWERS

7-1 (a) False. Natural selection acts upon random mutations.

(b) True

(c) False. Homologous proteins are derived from a common ancestor.

(d) True

(e) True

(f) False. Internal homology should be present in genes that evolved by contiguous gene duplication.

(g) True

(h) False. Generally, radical amino acid substitutions are deleterious to a protein's function.

(i) True

(j) False. The number of different residues is an underestimate due to silent mutations and multiple substitution at the same site.

(k) False. The Phe/Trp replacement is conservative but it requires two nucleotide substitutions and is found only rarely.

(l) True

(m) False. The net rate of fixation of neutral mutations in a population is equal to the mutation rate. The time required for fixation of a neutral mutation is $4N_e$ generations.

(n) True

(o) True

(p) False. Gene expansion and contraction occurs by recombination in misaligned regions that result from *homologous* pairing.

(q) True

(r) False. By definition, neutral mutations do not affect an organism's reproductive capacity, and consequently cannot be acted upon by natural selection.

(s) True

7-2 (a) Homologous

(b) Polyploidization

(c) Natural selection, genetic drift

(d) single-base substitution, gene duplication

(e) unequal crossing-over

(f) silent

(g) discrete

(h) DNA polymerase

(i) Conservative

(j) difference matrix

(k) Histone IV, fibrinopeptides

(l)   Orthologous, paralogous
(m)   genealogical (phylogenetic) tree
(n)   Coincidental evolution
(o)   gene expansion and contraction
(p)   X-ray crystallographic
(q)   satellite, multiplicational, informational

7-3   (a)   The four sets of homologous sequences are (1) and (5), which are from cytochrome *c;* (2) and (7), which are from α hemoglobin; (3), (6), and (8), which are from β hemoglobin; and (4), which is from fibrinopeptide A.
(b)   The homology between the α and β hemoglobin sets is more apparent after insertion of a gap between the first and second amino acids of the α hemoglobins.

7-4   The nucleotide difference matrix is given in Figure 7-27.

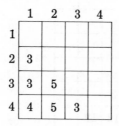

Figure 7-27.   Nucleotide difference matrix (Answer 7-4).

7-5   The most common amino acid at each position is assumed to be ancestral. This assumption leads to an ancestral sequence that could give rise to the contemporary sequences with the fewest genetic events.

```
               F
 I.  P A P C E Q G S A K K G A T L
           Y

                   V
 II. Y D L T Q P P S T S V S P G Q
                   Y

 III. V L S E G E W Q L V L H V W A
```

7-6   (a)   The best aligned sequences are

```
         1         5        10        15        20
    (1)  C S N L S T C V L S A Y W R N L N N F H
    (2)  C S [     ] T C V L S A Y W R N L N N F H
    (3)  [         ] Q C Y L S A Y W F N A N Q F H
    (4)  C S N L S T [         ] W R N I N Q F H
    (5)  [] S N L S T C L V S A Y W [     ] F H
    (6)  F T D V T S C L L [ ] F W K Q V [ ] Y R
```

(b)   The amino acid difference matrix can be derived directly from pairwise comparisons of the aligned sequences [Figure 7-28(a)]. Note that overlapping gaps are counted as one difference. The nucleotide difference matrix, which must be derived from a comparison of the aligned nucleotide sequences (Figure 7-29), is shown in Figure 7-28(b).

(c)   Sequence (3) differs from the others significantly more at the nucleotide level because of multiple two- and three-base substitutions.

(d)   The nucleotide comparison is a more accurate estimate of the genetic events that have separated these sequences because the multiple nucleotide changes (multiple genetic events) can be registered.

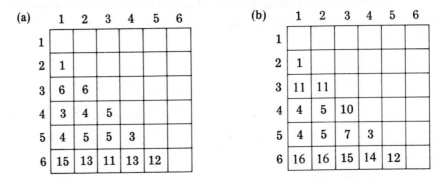

(a)

	1	2	3	4	5	6
1						
2	1					
3	6	6				
4	3	4	5			
5	4	5	5	3		
6	15	13	11	13	12	

(b)

	1	2	3	4	5	6
1						
2	1					
3	11	11				
4	4	5	10			
5	4	5	7	3		
6	16	16	15	14	12	

Figure 7-28. Difference matrices (Answer 7-6). (a) Amino acid differences. (b) Nucleotide differences.

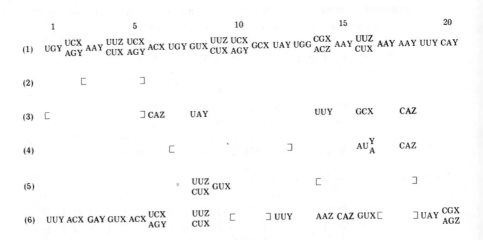

Figure 7-29. Nucleotide sequence alignments (Answer 7-6). Only the first complete sequence and the codons for the interchanged amino acids are shown.

7-7 (a) (1) If the generation times of these animals are very different, there will be more DNA replications, hence more opportunity for mutations to occur.

(2) Internal and external factors that affect mutation frequency might differ among these animals.

(3) The structure of the tree might be incorrect because of errors in the paleontological data.

(b) (1) The selection pressures on the different animals might be quite different.

(2) Some of the animals might have evolved primarily in small populations in which the founder principle might operate to fix larger numbers of mutations.

(3) Parallel and reverse mutations could distort symmetry. Reverse mutation will result in two nucleotide substitutions being counted as none. Since genealogical trees attempt to minimize the number of genetic events, there will be a tendency to view parallel mutations as a single genetic event that occurred prior to the divergence of the two species. In that case two mutational events will be counted as one.

(c) A genealogical tree should be more symmetrical if all fixed mutations are neutral, since the net rate of fixation by genetic drift depends only on the mutation rate. Fixation of non-neutral mutations depends on the forces of natural selection, which are not likely to remain identical over long time periods for all the organisms concerned.

7-8 (a) The α and β chains can be divided into three homology units 16 residues in length. The two β homology units differ from each other by 3 residues and from the α homology unit by 7 and 8 residues. Thus all homology units probably evolved from a primordial gene that coded for 16 residues. This gene duplicated to give ancestral α and β genes. The β gene underwent a contiguous gene duplication somewhat later and both genes diverged from one another to give contemporary α and β genes.

(b) The α chains may be coded by an informational gene family whose members exhibit conservative and highly restricted sequence diversity at each position. Perhaps these proteins carry out a variety of functions in the serum. Presumably all are the same size, and hence are isolated together by gel filtration even though only about 2% of these molecules exhibit the LSD-binding function. If this explanation is correct, the genes encoding these proteins may consitute a closely linked, multigene family with related functions.

7-9 (a) Since all the replacements in the variant hemoglobins can be accounted for by single-base substitutions, they support the contention that variant proteins arise by single-base substitutions.

(b) The amino acid replacements in the hemoglobin variants differ from those in evolutionarily related proteins in three ways.

(1) All hemoglobin variants can be accounted for by single-base substitutions.

(2)   Most hemoglobin variants involve changes in charge, whereas relatively few evolutionarily related proteins do.

(3)   There is a smaller proportion of conservative replacements in the hemoglobin variants than in the evolutionarily related proteins.

(c)   (1)   The evolutionarily related proteins presumably contain multiple-base substitutions because of multiple single-base substitutions over long evolutionary times.

        (2)   The hemoglobin variants contain a higher proportion of substitutions that involve charge changes because a large number of them were detected on that basis (difference in electrophoretic mobility).

        (3)   The smaller fraction of conservative replacements in the hemoglobin variants reflects the absence of natural selection.

7-10   These proteins may have arisen by a reiterative process in which a DNA polymerase copied a short stretch of DNA many times over to produce a gene consisting of repeated nucleotide sequences [e.g., (GCX ACX ACS)$_n$ for the antifreeze protein, where X is any nucleotide]. See M. Ycas, "*De novo* origin of periodic proteins," *J. Mol. Evol.* **2,** 17 (1972).

7-11   (a)   There are $3 \times 141 = 423$ nucleotides in the α chain. Since each nucleotide can be substituted by any of three other nucleotides, $423 \times 3 = 1269$ different nucleotide sequences can result from single-base substitutions.
(b)   The probability of a reverse mutation is 1/1269 and that of a forward mutation is 1268/1269. Thus a forward mutation is 1268 times as likely as a reverse mutation.
(c)   Parallel evolution will be unlikely in the absence of strong selection pressures to fix the very infrequent parallel changes much more efficiently than the far more frequent nonparallel substitutions.

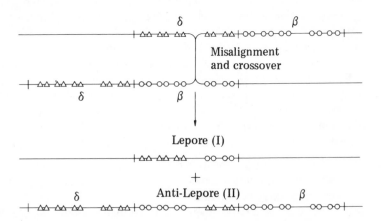

Figure 7-30.   Formation of hybrid genes Lepore and anti-Lepore (Answer 7-12).

7-12 (a) Variant I is identical with δ hemoglobin at its N terminus and with β hemoglobin at its C terminus, whereas Variant II shows just the reverse homology relationship. These properties of the two variants, considered together with the close linkage of the β and δ hemoglobin genes, suggest that Variants I and II arose by a crossover between misaligned β and δ genes (Figure 7-30). The recombination must have occurred in the region of the gene corresponding to residues 88 to 115. Variant I is hemoglobin Lepore and Variant II is hemoglobin anti-Lepore.

(b) The absence of normal β and δ hemoglobins in individuals with hemoglobin Lepore and their presence in individuals with hemoglobin anti-Lepore indicate that the order of the β and δ hemoglobin genes must be as shown in Figure 7-30.

7-13 (a) The α^2 allele contains an internal duplication (Residues 72-130 are a duplicate of Residues 13-71), which is the identifying mark of contiguous gene duplication. Since α^2 contains K at Position 54 and E at Position 113, it must have arisen from a recombination in a misaligned region between the genes for α^{1F} and α^{1S} as shown in Figure 7-31(a).

(b) Though it is possible that $\alpha^{2(KK)}$ and $\alpha^{2(EE)}$ arose independently by contiguous gene duplication as described for $\alpha^{2(KE)}$ in Part (a), it seems more likely

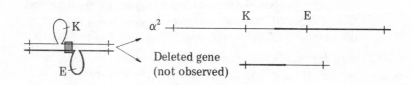

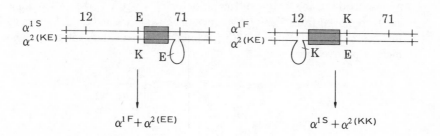

Figure 7-31. Production of various α^2 haptoglobin genes (Answer 7-13). (a) Contiguous gene duplication involving α^{1F} and α^{1S}. Recombination in the misaligned region (screened rectangle) produces α^2. (b) Conversion of $\alpha^{2(KE)}$ to $\alpha^{2(EE)}$ and $\alpha^{2(KK)}$. Recombination in the screened rectangles produces the indicated products. Note that the looped out portion of the chromosome, by virtue of the internal duplication, could be at any point between residues 12 and 71; only the two extreme positions are shown here.

that they arose by recombination in a homologous paired region between the genes for $\alpha^{2(KE)}$ and α^{1S} or α^{1F} as indicated in Figure 7-31(b).

(c)   Given the considerations of Part (b), it is impossible to decide which altered α chain arose first. As suggested in Part (a), $\alpha^{2(KE)}$ could have arisen after the one-base substitution. Alternatively, $\alpha^{2(KK)}$ could have arisen first and then been converted to $\alpha^{2(KE)}$ as described in Part (b) after α^{1S} arose.

7-14   (a)   The ratio of chain-termination (nonsense) mutations to total mutations as derived from the genetic code dictionary is 23/549 (Table 7-1). Therefore, $23/549 = 22/x$, in which x is the number of base substitutions occurring in the population you have studied $(x = 525)$. The fraction of possible base-substitution mutations that yield amino acid substitution (missense) mutations, according to the code, is 392/549. Therefore, $392/549 = y/525$, in which y is the number of amino acid substitutions in the population you studied $(y = 375)$.

(b)   The total number of mutants observed minus the number of mutants due to nonsense mutations equals the number of mutants that contain detectable amino acid substitutions $(65 - 22 = 43)$. Therefore, $43/375 \simeq 11\%$ of the estimated number of amino acid substitutions were detectable.

(c)   Presumably, the undetected mutants were missed because they carried mutations that were neutral or nearly neutral and did not seriously affect the function of the protein.

(d)   The number of silent mutations equals the total number of base substitutions expected minus the number of amino acid substitutions and the number of nonsense mutations expected $(525 - 375 - 22 = 128)$.

7-15   First construct a difference matrix to determine the relatedness of individual sequences (Figure 7-32). Sequences 1-2 and 3-4 are related most closely. These pairs are assigned to adjacent branches of a preliminary genealogical tree [Figure 7-33(a)]. The most probable ancestral sequences (5, 6, 7) are derived by inspection, using a simple "majority logic" (Figure 7-18). All ambiguities can be removed from 5 and 6 by comparison against 7 and a final genealogical tree can be drawn [Figure 7-33(b)]. Mutational events then can be assigned to their respective branches.

	1	2	3	4
1				
2	2			
3	3	3		
4	3	3	2	

Figure 7-32. Nucleotide difference matrix (Answer 7-15).

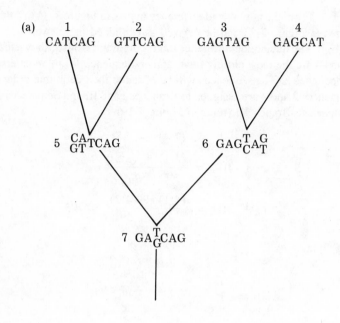

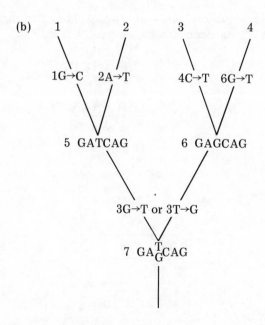

Figure 7-33. (a) Preliminary genealogical tree with ambiguous ancestral sequences. (b) Final genealogical tree with mutations assigned to appropriate branches (Answer 7-15).

7-16   (a)   From the amino acid difference matrix in Figure 7-34 (a), the genealogical tree in Figure 7-35 can be derived as explained in Answer 7-15.

(b)   The nucleotide difference matrix in Figure 7-34(b) shows that Sequences 2 and 4 are the most closely related, then Sequence 1, and most distantly related, Sequence 3. The sequences will be added to the tree in that order. Initially, Sequences 2 and 4 are assigned to Branches I and II, and Sequences 1 and 3 are assigned to Branch III (refer to Figure 7-19).

$$I \ + \ II \ = \ (2:4) \ = \ 2$$

$$I \ + \ III \ = \ \frac{(2:1) \ + \ (2:3)}{2} \ = \ 4.5$$

$$II \ + \ III \ = \ \frac{(4:1) \ + \ (4:3)}{2} \ = \ 4.5$$

Solving:

$$I \ = \ 1$$
$$II \ = \ 1$$
$$III \ = \ 3.5$$

In Step two, Sequences 2 and 4 are assigned to Branch I$'$, Sequence 1 is assigned to Branch II$'$, and Sequence 3 is assigned to Branch III$'$.

$$I' \ + \ II' \ = \ \frac{(2:1) \ + \ (4:1)}{2} \ = \ 4$$

$$I' \ + \ III' \ = \ \frac{(2:3) \ + \ (4:3)}{2} \ = \ 5$$

$$II' \ + \ III' \ = \ (1:3) \ = \ 3$$

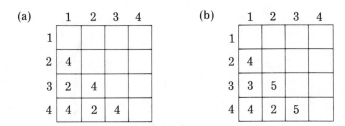

Figure 7-34. (a) Amino acid difference matrix. (b) Nucleotide difference matrix (Answer 7-16).

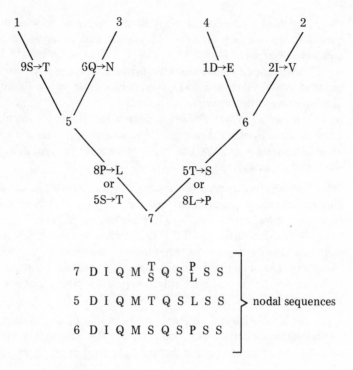

$$
\begin{array}{l}
7 \ D \ I \ Q \ M \ \overset{T}{S} \ Q \ S \ \overset{P}{L} \ S \ S \\[4pt]
5 \ D \ I \ Q \ M \ T \ Q \ S \ L \ S \ S \\[4pt]
6 \ D \ I \ Q \ M \ S \ Q \ S \ P \ S \ S
\end{array}
\Bigg\} \ \text{nodal sequences}
$$

Figure 7-35. Genealogical tree derived by the inspection method (Answer 7-16).

Solving:

$$
\begin{aligned}
I' &= 3 \\
II' &= 1 \\
III' &= 2
\end{aligned}
$$

From a comparison of these branch lengths the genealogical tree in Figure 7-36 can be derived.

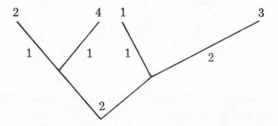

Figure 7-36. Genealogical tree derived by the Fitch-Margoliash method (Answer 7-16).

7-17   (a)   The third nucleotide of many codons can change without altering the encoded amino acid. Presumably, many of the third nucleotides in these two species have diverged.

(b)   The functional requirements for histone I V must be so precise that a ratio of one altered molecule to about a hundred normal molecules will confer a selective disadvantage on the organism.

(c)   Homogeneous, yet distinct, multigene families in the same species could be rationalized by either of two mechanisms for coincidental evolution: A variant arose in a master gene and all the copies were corrected against it, or the variant was fixed by crossover fixation.

7-18   (a)   The observed number of substitutions *(n)* first must be converted into the corrected number of substitutions *(m)* by Equation A7-1. Equation A7-2 then can be used to calculate the UEP's, which are shown in Table 7-8.

(b)   The different average UEP's for cytochrome *c* and globin probably result from much more stringent structure-function requirements for cytochrome *c* than for globin. Presumably the same number of mutational events are occurring in both families; however, natural selection permits fewer mutations to be fixed in the cytochrome *c* family.

7-19   (a)   The calculation of divergence time from UEP carries with it the implicit assumption that the rate of evolution is linear. This assumption probably is valid only for comparisons of genes that have identical or nearly identical functions

Table 7-8.   Corrected mutational differences per 100 residues, and unit evolutionary periods (UEP) for various related proteins (Answer 7-18). (From R. Dickerson, *J. Mol. Evol.* **1,** 26, 1971.)

	m	UEP (10^6 years)
Cytochrome *c*		
Birds/reptiles	13.6	17.7
Mammals/reptiles	15.3	19.6
Fish/other vertebrates	19.8	20.2
Vertebrates/insects	28.2	21.2
		Average 19.7
Globin		
Between β chains of nonprimates	24.8	3.6
Mammalian α/carp α	68.9	5.8
All α/all β, γ, δ	87.0	5.7
		Average 5.0

(e.g., cytochrome c). When a gene duplicates and one of the copies assumes a new function, the selective pressures will be different on the two genes. The gene with a new function will change more rapidly as it evolves a structure more suitable to the new function.

(b)   The calculated value is an overestimate (by at least 2.0×10^8 years), since hemoglobin, the gene with the new function, has evolved at a more rapid rate than myoglobin.

(c)   Since these different hemoglobin chains do have some distinct functions, the resulting UEP is to some extent an approximation.

7-20   (a)   Sequence (1) is a portion of the α chain of haptoglobin (Residues 22-38) and Sequence (2) is a portion of an immunoglobulin light chain (Residues 73-89).

$$AMV = \frac{19 \text{ substitutions}}{17 \text{ codons}} = 1.12$$

These sequences are probably homologous, although certainly on the border line.

(b)   Sequence (1) is the amino-terminal portion of lysozyme and Sequence (2) is the corresponding region of ribonuclease.

$$AMV = \frac{17 \text{ substitutions}}{12 \text{ codons}} = 1.41$$

These sequences are not homologous.

(c)   Sequence (1) is the amino-terminal sequence of an immunoglobulin heavy chain and Sequence (2) is the corresponding region of a light chain.

$$AMV = \frac{23 \text{ substitutions}}{22 \text{ codons}} = 1.05$$

These sequences are probably homologous.

7-21   The amino acid difference matrix for these proteins is given in Figure 7-37 and

	S1	S2	S3	S4	S5	S6	S7	S8
S1								
S2	15							
S3	10	14						
S4	10	14	3					
S5	14	5	13	13				
S6	9	14	1	2	13			
S7	10	15	3	2	14	2		
S8	15	0	14	14	5	14	15	

Figure 7-37.   An amino acid difference matrix (Answer 7-21).

the derived genealogical tree is shown in Figure 7-38. The amino acid difference matrix indicates that S3, S4, S6, and S7 are closely related, as are S2, S5, and S8. S1 is more closely related to the S3 cluster than to the S2 cluster. Nodal sequence *a* can be derived by a comparison of S3, S4, S6, and S7. Nodal sequence *b* was derived from S2, S5, and S8 and the ambiguities were resolved by comparison with S1 and nodal sequence *a*. Nodal sequence *b* was derived from a comparison of S1 and nodal sequences *a* and *c*. Nodal sequence *d* was derived from nodal sequences *b* and *c*. The amino acid substitutions that occurred in each branch were determined by comparison of the individual sequence with the appropriate nodal sequence.

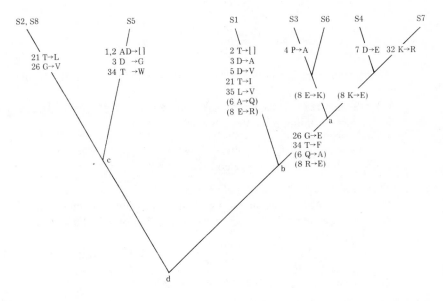

Nodal sequences

Figure 7-38.   A genealogical tree for proteins S1 to S8 (Answer 7-21). Parentheses indicate amino acid substitutions that cannot be assigned uniquely to a branch.

7-22 (a) Assuming that every 1.5% mismatch produces a 1°C drop in T_m, the percent nucleotide divergence between green monkey and man is (7°C) (1.5%/°C) or 10.5% and between green monkey and capuchin is (12.3°C) (1.5%/°C) or 18.5%.

(b) The corrected nucleotide divergences (from Equation A7-1) are 11.2% and 20.3%. The UEP's for unique DNA from green monkey and man and from green monkey and capuchin, respectively, are 4.0×10^6 years and 3.2×10^6 years.

(c) The rate of nucleotide change for rodents is more than 10 times that of the primates. Apparently, these differences are not linear with regard to absolute time. If the rate of nucleotide change is related to generation time and not to absolute time, rodents would be expected to evolve faster than primates, since they have a much shorter generation time.

7-23 (a) Mouse satellite DNA appears to be composed of repeating sequences that contain the Eco RII cleavage site. Therefore partial digestion with Eco RII leads to a major series of bands that represent monomers, dimers, trimers, and so on, of the repeating sequence. The minor bands suggest that the satellite contains a few additional cleavage sites near the centers of the intervals between primary cleavage sites, so that a few 1/2-mers, 1 1/2-mers, 2 1/2-mers, and so on, also are produced.

(b) 3840 nucleotide pairs/16 = 240 nucleotide pairs.

(c) Assuming that 1.5% mismatch produces a 1°C drop in T_m, these sequences show about 7.5% divergence.

(d) The formation of polymers and circles in the monomer fraction indicates that the repeat unit for reassociation is less than 240 nucleotides. Consideration of the fraction of monomers, dimers, and trimers that reassociate as linears suggests a repeat unit for reassociation of 120 nucleotides. As indicated in Figure 7-39, monomers have a 50% chance of reassociation in register and dimers have

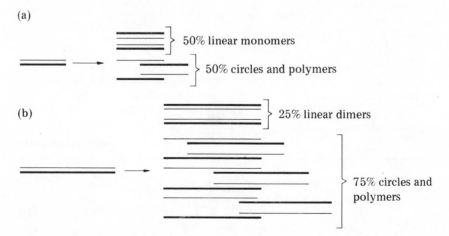

Figure 7-39. Registration of restriction fragments after reassociation (Answer 7-23). (a) Monomers. (b) Dimers. (Adapted from E. M. Southern, *J. Mol. Biol.*, in press, 1975.)

a 25% chance. Trimers have a 16 2/3% chance of reassociation in register. These theoretical figures agree very well with the actual reassociation experiments.

(e)   As indicated in Figure 7-40, unequal crossing-over when the DNA is misaligned in the 120 nucleotide register will produce a 1/2-mer and a 1 1/2-mer.

(f)   One plausible scheme for the evolution of mouse satellite DNA is shown in Figure 7-41.

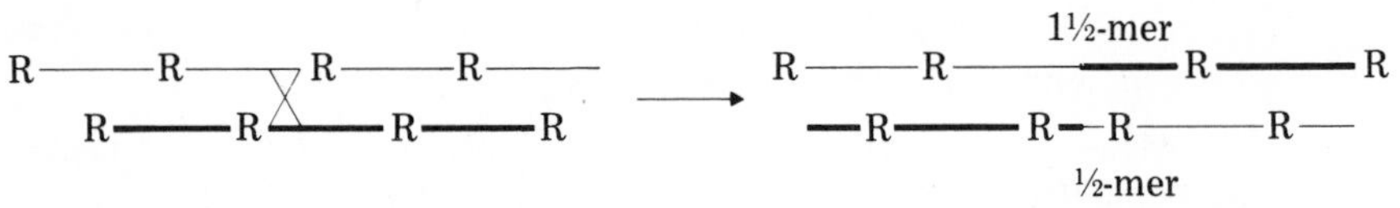

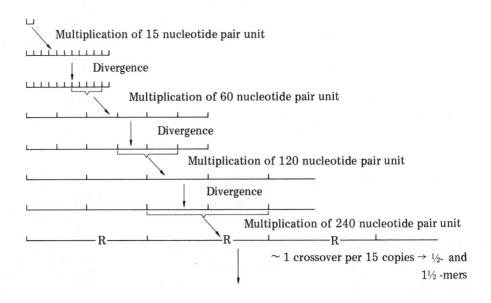

Figure 7-40.   A scheme for production of 1/2-mers and 1 1/2-mers by unequal crossing-over (Answer 7-23). (Adapted from E. M. Southern, *J. Mol. Biol.,* in press, 1975.)

Figure 7-41.   A scheme for the evolution of mouse satellite DNA (Answer 7-23.) (Adapted from E. M. Southern, *J. Mol. Biol.,* in press, 1975.)

7-24 Inspection of the nucleotide difference matrix for the globins shows that β and δ hemoglobin are most closely related. γ hemoglobin is next most closely related, then α hemoglobin, and finally myoglobin. In the Fitch-Margoliash method the globins are added to the growing tree in that order. The construction of the tree is as follows:

Step 1. [See Figure 7-42(a).]

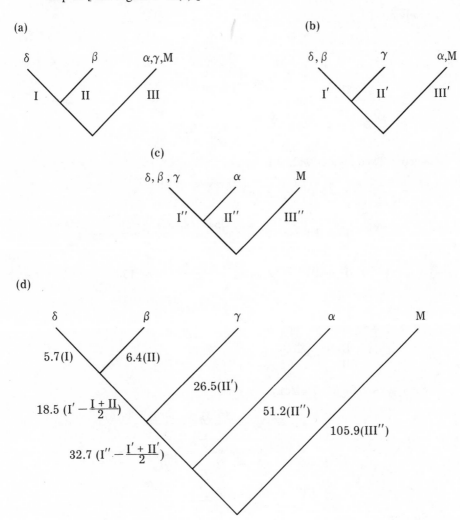

Figure 7-42. Construction of a genealogical tree for globin genes (Answer 7-24). (a), (b), and (c) show the intermediate trees used in the Fitch-Margoliash method. (d) is the genealogical tree for the globin genes. Mutational distances are indicated on appropriate branches.

$$I + II = X = (\delta:\beta) = 12$$

$$I + III = Y = \frac{(\delta:\alpha) + (\delta:\gamma) + (\delta:M)}{3} = 107.3$$

$$II + III = Z = \frac{(\beta:\alpha) + (\beta:\gamma) + (\beta:M)}{3} = 108$$

Solving:

$$I = \frac{X + Y - Z}{2} = 5.7$$

$$II = \frac{X - Y + Z}{2} = 6.4$$

$$III = \frac{-X + Y + Z}{2} = 101.7$$

Step 2. [See Figure 7-42(b).]

$$I' + II' = X = \frac{(\delta:\gamma) + (\beta:\gamma)}{2} = 51$$

$$I' + III' = Y = \frac{(\delta:\alpha) + (\delta:M) + (\beta:\alpha) + (\beta:M)}{4} = 136$$

$$II' + III' = Z = \frac{(\gamma:\alpha) + (\delta:M)}{2} = 138$$

Solving:

$$I' = 24.5$$
$$II' = 26.5$$
$$III' = 111.5$$

Step 3. [See Figure 7-42(c).]

$$I'' + II'' = X = \frac{(\delta:\alpha) + (\beta:\alpha) + (\gamma:\alpha)}{3} = 109.3$$

$$I'' + III'' = Y = \frac{(\delta:M) + (\beta:M) + (\gamma:M)}{3} = 164$$

$$II'' + III'' = Z = (\alpha:M) = 157$$

Solving:

$$I'' = 58.2$$
$$II'' = 51.2$$
$$III'' = 105.9$$

These mutation distances are related to the final genealogical tree for the globins as indicated in Figure 7-42(d).

INDEX

This index is designed to reflect the organization of the book. Thus, the index does not refer to page numbers, rather, the following abbreviations are used: E = Essential Concept, EA = Appendix to Essential Concepts, F = Figure, T = Table, and PA = Problem and Answer. For example, E3-3 refers to Essential Concept 3-3, PA3-3 refers to Problem and Answer 3-3, T3-3 refers to Table 3-3, and so on.

Hemoglobin (continued)
  mRNA, E2-13(a), E5-12, T5-1
  unit evolutionary period, PA7-18
Hemolymph, E5-9
Hepatoma, E4-9, PA4-17
Hermaphrodite, PA3-26
Heterochromatin
  chromosome loss, PA4-15
  chromosome structure, E2-7
  control of gene expression, T5-1
  satellite DNA, E2-11(b), PA2-15
Heteroduplex
  gene conversion, E3-8(d), F3-10
  recombination, E3-7(b), F3-8
Heterogeneous nuclear RNA (see
    hnRNA)
Heterokaryon
  analysis, E4-7
  cell fusion, E4-6(b), F4-5
  macromolecular synthesis, PA4-8,
    PA4-14
Heterozygous, E3-1(a)
Hexoseaminidase
  chromosome assignment, PA4-10
Hfr, E1-1(c), F1-1
HGPRT
  chromosome assignment, PA4-11
  hybrid selection, E4-6(e)
  Lesch-Nyhan syndrome, PA4-6
Highly repetitive DNA (see
    Satellite DNA)
Hippo hypotheticus, PA3-15,
    PA3-16, PA3-17, PA3-18
Histidine operon, E1-6(b), F1-4
Histone
  amino acid replacements, E2-5(a),
    E7-6(e)
  chromatin, E2-3, PA5-12
  chromatin reconstitution, PA6-3
  classes, E2-5(a), T2-1
  function, PA2-14
  gene localization, E2-14(b)
  histone IV evolutionary rate, T7-2
  histone IV sequence, F2-19, PA2-14
  multigene family, PA7-17
  phosphorylation, E6-7(c)
  structural unit of chromatin, E2-6
hnRNA
  characteristics, E5-11
  poly-A, E5-12, PA5-11
Homoeotic mutants
  Britten-Davidson model, E5-15
  Drosophila, E5-5(b), F5-7
Homokaryon, E4-6(b)
Homologous protein
  sequence comparisons, EA7-1,
    E7-7, PA7-3, PA7-5, PA7-6,
    PA7-20
  statistical comparisons, PA7-20
Homologue, E2-2(a), F2-1

Homozygous, E3-1(a)
Hormone
  adrenocorticotropic hormone,
    E6-2(a), F6-1, T6-1
  aldosterone, F6-1, F6-3, T6-1
  Britten-Davidson model, E5-15
  calcitonin, F6-1, T6-1
  cAMP, E6-6, E6-7, F6-5, PA6-11,
    T6-2
  chemical classes, E6-2
  cholecystokinin, F6-1, T6-1
  circulating concentration, PA6-15
  corticosterone, F6-1, F6-3, T6-1
  cortisol, E6-3, F6-1, F6-3, T6-1
  developmental, E6-4, PA6-5
  ecdysone, E2-8(e), E5-9, F5-9
  (see Epinephrine)
  erythropoietin, F5-2
  (see Estradiol)
  follicle-stimulating hormone, F6-1,
    T6-1
  gastrin, F6-1, T6-1
  glucagon, E6-2(a), E6-7(c), F6-1,
    T6-1
  growth hormone, E6-2(a), F6-1,
    T6-1
  hydrocortisone, PA5-13, PA6-5
  (see Insulin)
  lutenizing hormone, F6-1, T6-1
  norepinephrine, E6-1(b), E6-2(a),
    F6-1, F6-2, T6-1
  oxytocin, F6-1, T6-1
  parathormone, E6-2(a), F6-1, T6-1
  physiology, E6-1, F6-1, T6-1
  post-translational modification,
    E5-14(a)
  (see Progesterone)
  prolactin, F6-1, T6-1
  (see Prostaglandins)
  protein synthesis, E6-8
  (see Receptor)
  secretin, F6-1, T6-1
  secretion and inactivation, E6-3
  somatomedin, F6-1, T6-1
  somatotropin, F6-1, T6-1
  testosterone, F6-1, F6-3, T6-1
  thyrotropic hormone, F6-1, T6-1
  thyroxine, E6-2(a), E6-3, F6-1,
    T6-1
  vasopressin, F6-1, T6-1
Household
  cell functions, E4-9
Human
  amniocentesis, E4-11(a)
  cell number, E3-2(b)
  cloning, E4-11(b)
  differentiated tumor, E4-9
  DNA content, E2-4(b), F2-7
  gene mapping, E4-8, PA4-9,
    PA4-10, PA4-11

*The explanation of the index is found on Page 325